Grade 5

21st Century Assessments

www.mheonline.com

 McGraw-Hill is committed to providing Instruction materials in Science, Technology, Engineering, and Mathematics (STEM) that give all students a solid foundation, one that prepares them for college and careers in the 21st century.

Send all inquiries to:
McGraw-Hill Education
8787 Orion Place
Columbus, OH 43240

ISBN: 978-0-07-667440-4
MHID: 0-07-667440-1

Printed in the United States of America.

3 4 5 6 7 8 9 10 LHS 22 21 20 19 18

STEM Our mission is to provide educational resources that enable students to become the problem solvers of the 21st century and inspire them to explore careers within Science, Technology, Engineering, and Mathematics (STEM) related fields

Contents

Teacher's Guide to 21st Century Assessment Preparation

Whether it is the print *21st Century Assessments* or online at **ConnectED**. mcgraw-hill.com, *McGraw-Hill My Math* helps students prepare for online testing.

How to Use this Book

21st Century Assessments includes experiences needed to prepare students for the upcoming online state assessments. The exercises in this book give students a taste of the different types of questions that may appear on these test.

Assessment Item Types

- Familiarizes students with commonly-seen item types

- Each type comes with a description of the online experience, helpful, hints, and a problem for students to try on their own.

Countdown

- Prepares students in the 20 weeks leading up to online assessments

- Consists of five problems per week, paced with order of the *McGraw-Hill My Math Student Edition* with built in review.

- **Ideas for Use** Begin use in October for pacing up to the beginning of March. Assign each weekly countdown as in-class work for small groups, homework, a practice assessment, or a weekly quiz. You may assign one problem per day or have students complete all five problems at once.

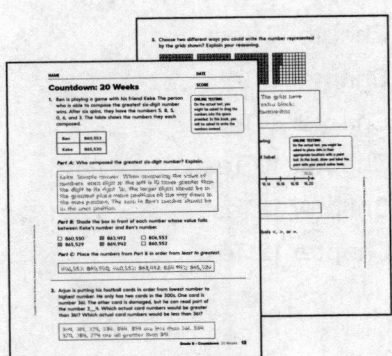

Chapter Tests

- Each six-page test assesses all of the standards for mathematics presented in the chapter.

- Each question mirrors an item type that might be found on online assessments, including multi-part questions.

- **Ideas for Use** Assign as in-class group work, homework, a practice assessment, a diagnostic assessment before beginning the chapter, or a summative assessment upon completing the chapter.

Chapter Performance Tasks

- Each two-page performance task measures students' abilities to integrate knowledge and skills across multiple standards. This helps students prepare for the rigor expected in college and future careers.

- A rubric describes the standards assessed and guidelines for scoring student work for full and partial credit.

- Sample student work is also included in the answer section of this book.

- **Ideas for Use** Assign as in-class small group work, homework, a practice assessment, or in conjunction with the Chapter Test as part of the summative assessment upon completion of the chapter.

Benchmark Tests

Four benchmark tests are included in this book. All problems on the tests mirror the item types that may be found on online assessments. Each benchmark test also includes a performance task.

- The *first* benchmark test is an eight-page assessment that addresses the standards from the first third of the Student Edition.

- The *second* benchmark test addresses the second third of the Student Edition.

- The *third* and *fourth* benchmark tests (Forms A and B) are twelve-page assessments that address the standards from the entire year, all chapters of the Student Edition.

- A rubric is provided in the Answer section for scoring the performance task portion of each test.

- **Ideas for Use** Each benchmark test can be used as a diagnostic assessment prior to instruction or as a summative assessment upon completion of instruction. Forms A and B can be used as a pretest at the beginning of the year and then as a posttest at the end of the year to measure mastery progress.

Go Online for More! connectED.mcgraw-hill.com

Performance Task rubrics to help students guide their responses are also available. These describe the tasks students should perform correctly in order to receive maximum credit.

Additional year-end performance tasks are available for Grades K through 5 as blackline masters available under Assessment in ConnectED.

Students can also be assigned tech-enhanced questions from the eAssessment Suite in ConnectED. These questions provide not only rigor, but the functionality students may experience when taking the online assessment.

Assessment Item Types

In the spring, you will probably take a state test for math that is given on a computer. The problems on the next few pages show you the kinds of questions you might have to answer and what to do to show your answer on the computer.

Selected Response means that you are given answers from which you can choose.

Selected Response Items

Regular multiple choice questions are like tests you may have taken before. Read the question and then choose the <u>one</u> best answer.

Multiple Choice

Four yards of fabric will be cut into pieces so that each piece is thirteen inches long. How many pieces can be cut?

☐ 6 pieces with 2 inches left over

☐ 7 pieces with I inch left over

☐ 10 pieces with 2 inches left over

☐ II pieces with I inch left over

ONLINE EXPERIENCE Click on the box to select the one correct answer.

HELPFUL HINT Only one answer is correct. You may be able to rule out some of the answer choices because they are unreasonable.

▶ Try On Your Own!

Four boxes to be mailed are weighed at the post office. Box A weighs 8.22 pounds, Box B weighs 8.25 pounds, and Box C weighs 8.225 pounds. Box D weighs less than Box C but more than Box A. How much could Box D weigh?

☐ 8.22 pounds

☐ 8.224 pounds

☐ 8.226 pounds

☐ 8.23 pounds

Sometimes a multiple choice question may have more than one answer that is correct. The question may or may not tell you how many to choose.

Multiple Correct Answers

Select **all** values that are equivalent to 332 ounces.

- ☐ 2 gallons, 76 ounces
- ☐ 20 pints, 12 ounces
- ☐ 22 pints, 8 ounces
- ☐ 41 cups, 5 ounces
- ☐ 41 cups, 4 ounces

ONLINE EXPERIENCE Click on the box to select it.

HELPFUL HINT Read each answer choice carefully. There may be more than one right answer.

Try On Your Own!

Select **all** statements that are true.

- ☐ All rhombuses are parallelograms.
- ☐ All trapezoids are parallelograms.
- ☐ All rectangles are trapezoids.
- ☐ All squares are rectangles.

Another type of question asks you to tell whether the sentence given is true or false. It may also ask you whether you agree with the statement, or if it is true. Then you select yes or no to tell whether you agree.

Determine whether each polygon shown is also a rhombus. Select Yes or No for each polygon.

ONLINE EXPERIENCE Click on the box to select it.

HELPFUL HINT There is more than one statement. Any or all of them may be correct.

Yes   No

☐   ☐

☐   ☐

☐   ☐

☐   ☐

▶ **Try On Your Own!**

Select True or False for each comparison.

True   False

☐   ☐   200 centimeters > 1.5 meters

☐   ☐   36 inches > 2 yards

☐   ☐   1 gallon > 12 cups

☐   ☐   2 miles < 3,500 yards

You may have to choose your answer from a group of objects.

Click to Select

A rectangular prism has a length of 12 centimeters, a width of 8 centimeters, and a height of 32 centimeters. Which equations could be used to find the volume of the rectangular prism in cubic centimeters?

$12 + 8 + 32 = V$ $12 \times 8 \times 32 = V$

$(12 + 8) \times 32 = V$ $(32 \times 8) \times 12 = V$

$96 \times 32 = V$ $18 \times 32 = V$

ONLINE EXPERIENCE
Click on the figure to select it.

HELPFUL HINT On this page you can draw a circle or a box around the figure you want to choose.

Try On Your Own!

Select **all** expressions that are equal to $5\frac{1}{3}$.

$16 \times \frac{1}{3}$ $2\frac{1}{3} \times 2\frac{2}{3}$ $32 \times \frac{1}{6}$

$15 \times \frac{1}{3}$ $8 \times \frac{2}{3}$ $3\frac{1}{3} \times 2$

When no choices are given from which you can choose, you must create the correct answer. One way is to type in the correct answer. Another may be to make the correct answer from parts that are given to you.

Constructed-Response Items

Fill in the Blank

The table shows the number of laps Tammi ran around the track each day. Complete the table if the pattern continues.

Day	Laps
1	4
2	7
3	10
4	
5	
6	
7	

ⓘ ONLINE EXPERIENCE You will click on the space and a keyboard will appear for you to use to write the numbers and symbols you need.

💡 HELPFUL HINT Be sure to provide an answer for each space in the table.

Try On Your Own!

Sasha planted a garden in her backyard that is 32 square feet in area. If the length was 8 feet, how many inches wide was the garden?

Sometimes you must use your mouse to click on an object and drag it to the correct place to create your answer.

Drag one expression to each box to make the statements true.

Subtract 3 from 9 and then add 2.	=	

Add 3 and 9 and then subtract 2.	=	

The sum of 3 and 2 is subtracted from 9.	=	

ONLINE EXPERIENCE
You will click on an expression and drag it to the spot it belongs.

HELPFUL HINT Either draw a line to show where the expression goes or write the expression in the blank.

$$3 + 9 - 2 \qquad 9 - (3 + 2) \qquad 9 - 3 + 2$$

▶ **Try On Your Own!**

Order from least to greatest by dragging each number to a box.

3.045	3.109	3.103	3.17	3.016	3.059

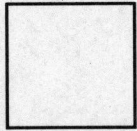 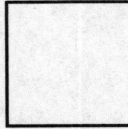 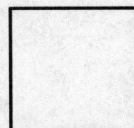 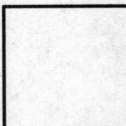 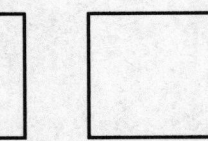

Some questions have two or more parts to answer. Each part might be a different type of question.

Multipart Question

Connor is filling a 15-gallon wading pool with water. On his first trip he carried $3\frac{7}{12}$ gallons of water. He carried $3\frac{1}{3}$ gallons on his second trip, and $2\frac{1}{2}$ gallons on his third trip.

Part A: How much water did Connor carry to the wading pool on trips 1, 2, and 3?

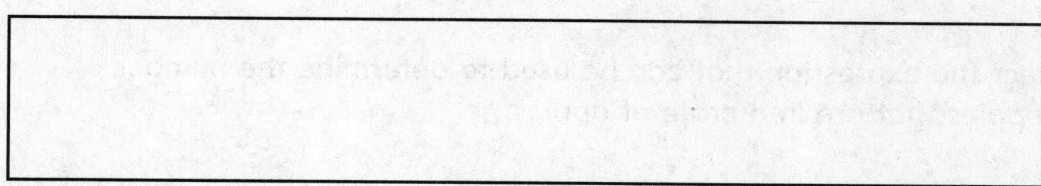

Part B: How many more gallons will Connor need to carry to the wading pool until it is filled?

Try On Your Own!

This table shows the three different ways that apples are sold at Donaldson's Fruit Farm in the fall.

Package Type	Amount in the Package
Bag	12 apples
Box	8 bags
Crate	15 boxes

Part A: Select the expression that can be used to determine the number of bags of apples that are in a crate of apples.

☐   8 × 15

☐   12 + 8 + 15

☐   12 × 8 × 15

☐   12 × 8

Part B: The label on a bag of apples states that it contains 1.25 pounds of apples. What is the total weight, in pounds, of the bags of apples in one crate?

One crate of apples weighs ⬚ pounds.

Countdown: 20 Weeks

1. Ben is playing a game with his friend Keke. The person who is able to compose the greatest six-digit number wins. After six spins, they have the numbers 5, 8, 5, 0, 6, and 3. The table shows the numbers they each composed.

| Ben | 860,553 |
| Keke | 865,530 |

Part A: Who composed the greatest six-digit number? Explain.

Part B: Shade the box in front of each number whose value falls between Keke's number and Ben's number.

☐ 860,550   ☐ 863,492   ☐ 806,553
☐ 865,529   ☐ 864,942   ☐ 860,552

Part C: Place the numbers from Part B in order from *least to greatest*.

2. Arjun is putting his football cards in order from lowest number to highest number. He only has two cards in the 300s. One card is number 361. The other card is damaged, but he can read part of the number 3__4. Which actual card numbers would be greater than 361? Which actual card numbers would be less than 361?

3. Choose two different ways you could write the number represented by the grids shown? Explain your reasoning.

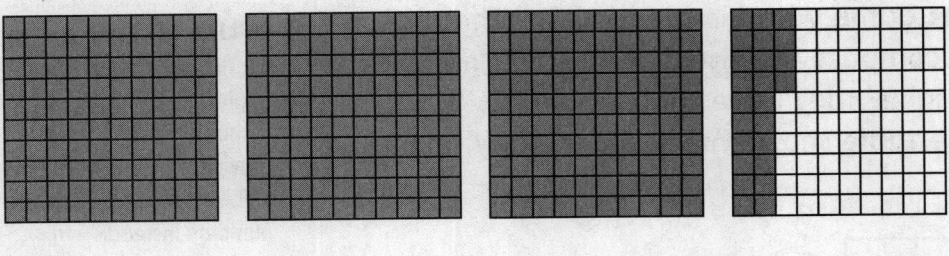

4. Triple jumps at a track meet have the following distances: 16.08 m, 16.1 m, 16.02 m, 16.20 m.

Part A: Place a dot on the number line and label for each given distance.

ONLINE TESTING
On the actual test, you might be asked to place dots in their appropriate locations with a point tool. In this book, draw and label the point with your pencil online tools.

Part B: Which jump was the longest?

5. Compare each number to 2.15. Use the symbols <, >, or =.

2.150 ◯ 2.15

$2 + \frac{15}{10}$ ◯ 2.15

2 ones and 15 thousandths ◯ 2.15

215 hundredths ◯ 2.15

Countdown: 19 Weeks

1. Julian was taking notes for a report on the U.S. population. When reading his notes later, he found he couldn't read all the numbers. He did remember the following information.

A. The smallest place-value position is 6.

B. The number in the hundred thousands place has a value that is $\frac{1}{10}$ the value of the number in the millions place.

C. The value of the number 7 is 7 thousands.

Using the hints from above, write the missing digits in the chart.

3	1		8	5		0	5	

2. Julian compared numbers with similar digits. Using mathematical language, explain how each set of numbers is different.

A	53,671 and 52,671
B	354 and 3.54
C	152 and 0152

3. There are 10 years in every decade, 100 years in every century, and 1000 years in every millenium.

Part A: How many decades are there in a millenium? Explain.

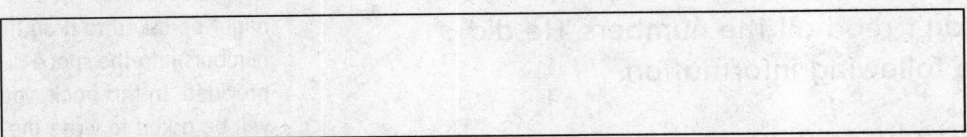

Part B: How many centuries are there in a millenium? Explain.

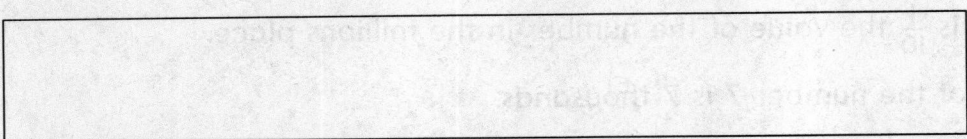

4. Compare $\frac{3}{10}$ and $\frac{3}{100}$.

Part A: Shade the decimal models to show each fraction.

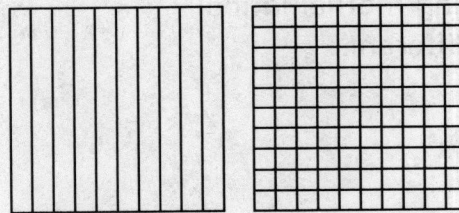

Part B: Compare the two decimals. Use >, <, or =. Explain.

5. How can you use models to explain why 3.1 = 3.10?

Countdown: 18 Weeks

1. Find the prime factorization for the number below.

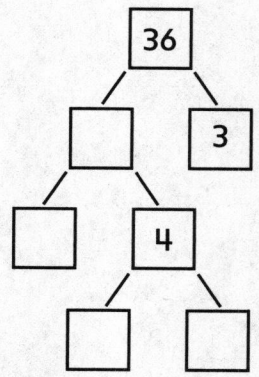

36 = _____

2. Each box of paper clips contains 10^3 clips. The school store has 25 boxes.

   Part A: What is the value of 10^3?

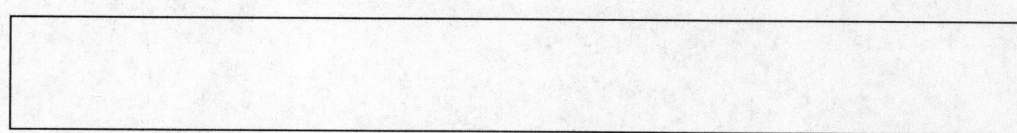

   Part B: How many paper clips does the school store have?

3. A recipe for pancakes calls for 3 cups of flour for every 2 tablespoons of sugar. Fill in the chart to find how many cups of flour are needed for 8 tablespoons of sugar.

Tablespoons of Sugar	Cups of Flour
2	3
4	
6	
8	

4. Find 8 × 52 using an area model.

```
           50                    2
      ┌──────────────────┬──────────────────┐
  8   │  ☐ × ☐           │   ☐ × ☐          │
      └──────────────────┴──────────────────┘
```

$8 \times 52 = (8 \times \quad) + (8 \times \quad)$

$= \quad + \quad$

$= \quad$

5. Shade the box in front of the statements that are true.

☐ 3.240 > 3.24

☐ 2 and 34 hundredths = 2.34

☐ $2 + \frac{3}{10} = 2.3$

☐ 536 hundredths = 53.6

Countdown: 17 Weeks

1. ABC Pens sells pens in boxes of 12. Their competitor XYZ Pens sells pens in boxes of 144. An office building is considering purchasing either 10^3 boxes from ABC Pens or 10^2 boxes from XYZ Pens.

Part A: How many pens are in 10^3 boxes of ABC Pens?

Part B: How many pens are in 10^2 boxes of XYZ Pens?

2. Circle the problems that have a correct solution.

$$\begin{array}{r} 254 \\ \times\ 12 \\ \hline 3{,}048 \end{array}$$

$$\begin{array}{r} 412 \\ \times\ 24 \\ \hline 9{,}788 \end{array}$$

$$\begin{array}{r} 316 \\ \times\ 29 \\ \hline 9{,}164 \end{array}$$

$$\begin{array}{r} 581 \\ \times\ 32 \\ \hline 18{,}592 \end{array}$$

ONLINE TESTING
On the actual test, you might be asked to click on the problem to put a circle around it. In this book, you will be asked to make the circles with a pencil instead.

3. The table below lists the number of students in each grade level of an elementary school. Estimate how many students are in the school by rounding. Show how you estimated.

Grade Level	Number of Students
Kindergarten	315
First Grade	378
Second Grade	412
Third Grade	351
Fourth Grade	401
Fifth Grade	345

4. A penny is 1.52 mm thick. Write this number in expanded form.

5. The land area of Arizona is $(1 \times 100,000) + (1 \times 10,000) + (3 \times 1,000) + (9 \times 100) + (9 \times 10) + (8 \times 1)$ square miles.

Part A: Write the correct digits in the boxes in order to put the number into standard form.

☐ ☐ ☐ , ☐ ☐ ☐

Part B: Write the area of Arizona in words.

Countdown: 16 Weeks

1. For the numbers 6, 7, and 42, circle the equations that are members of the fact family.

$6 \times 7 = 42$ $42 \div 7 = 6$

$7 + 6 = 13$ $7 \times 6 = 42$

$7 \times 42 = 6$ $42 \div 6 = 7$

2. A candy company puts 200 pieces of candy inside the bag. In the month of July, the company sold 8,000,000 pieces of candy. Determine whether each statement will find the number of bags of candy the company sold in July.

Yes   No

☐   ☐ $8{,}000{,}000 \div 200$

☐   ☐ $800{,}000 \div 200$

☐   ☐ $8{,}000{,}000 \div 20$

☐   ☐ $800{,}000 \div 20$

☐   ☐ $80{,}000 \div 2$

> **ONLINE TESTING**
> On the actual test, you might be asked to click on a box to shade it. In this book, you will be asked to shade the box with a pencil instead.

3. The table shows the amount that a painter charges for painting rooms. If your house has four bedrooms, two bathrooms, and three other rooms, how much will it cost to have the entire house painted?

Type of Room	Cost
Bedroom	$100
Bathroom	$50
Other Rooms	$120

4. A roller coaster can take 24 riders in a single trip. 72 people went through the line to ride the roller coaster.

Part A: How many trips did the roller coaster make?

>

Part B: Write the multiplication and division fact family for this.

>

---

5. *Part A:* A school building has 67 classrooms in it. Four students have volunteered to clean the classrooms over summer break. Fill in the boxes to find how many rooms each student should clean.

$$
\begin{array}{r}
\square\,\square\ \text{R}\,\square \\
4\overline{)6\ 7} \\
-\square \\
\hline
\square\,\square \\
-\square\,\square \\
\hline
\square
\end{array}
$$

Each student should clean _____ rooms.

Part B: The building principal has offered to clean the left over rooms. How many rooms will she clean?

>

Countdown: 15 Weeks

1. **Part A:** Eggs are sold by the dozen. If a chicken farm has produced 2,386 eggs, color in the box next to any expression that will estimate how many dozens can be packaged.

   ☐ 2,000 ÷ 12

   ☐ 2,400 ÷ 12

   ☐ 3,000 ÷ 12

   ☐ 2,300 ÷ 12

   Part B: Estimate the number of dozens that can be packaged.

   ┌──┐
   │   │
   │   │
   └──┘

2. Jerry and his two friends are going to bake cookies for a fundraiser. They need to bake 369 cookies in all. Use a model to find the number of cookies each person needs to bake.

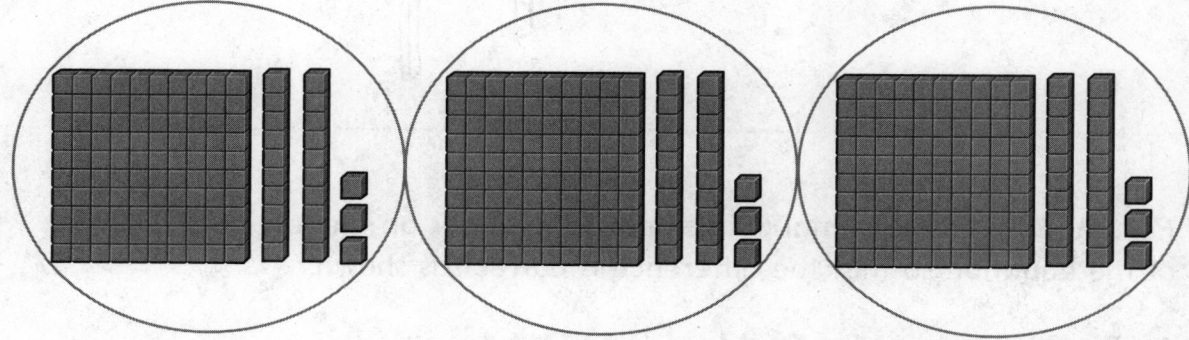

   Each person bakes _____ cookies.

> **ONLINE TESTING**
> On the actual test, you might be asked drag and drop groups of hundreds, tens, and ones to make the model. In this book, you will be asked to make the model by drawing instead.

3. Use the Distributive Property to draw a bar diagram and solve the problem.

$936 \div 3 =$ _____

---

4. Chairs for a dining room set cost $78 each. Circle the equation that can be used to best estimate the cost of buying chairs for a family of 6.

$70 \times 6 = \$420$

$80 \times 6 = \$480$

$100 \times 6 = \$600$

$50 \times 6 = \$300$

---

5. *Part A:* Draw the decimal points on each number on the left side of the equation so that the difference is correct as shown.

$1\ 2\ 3\ 4 - 6\ 2\ 4 = 117.16$

Part B: Check to make sure that the answer is reasonable by rounding.

Countdown: 14 Weeks

1. There are 653 rubber bands in a desk drawer. The teacher wants to split them as evenly as possible among 63 students. Circle the equation that is the least accurate estimate for the number of rubber bands each student should receive.

$$660 \div 66 = 10$$

$$650 \div 65 = 10$$

$$650 \div 50 = 13$$

$$480 \div 60 = 8$$

> **ONLINE TESTING**
> On the actual test, you might be asked click to draw a circle. In this book, you will be asked to draw the circle using a pencil.

2. Ms. Chen wants to purchase sets of Christmas lights to decorate her house. The lights cost $13 per package. She has saved $360 for the project.

Part A: How many packages of lights can she buy with $360?

Part B: What is the remainder, and what does it represent?

Part C: Round to estimate the answer so you can check for reasonableness.

3. A farmer has a rectangular field to plow. The field has an area of 18,963 square yards. The field is shown below. Fill in the missing length.

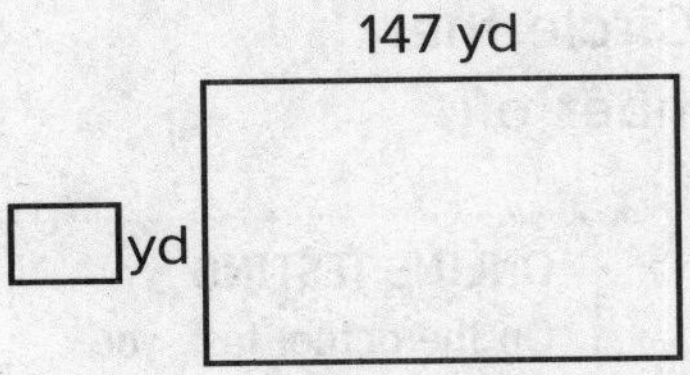

147 yd

☐ yd

- -

4. Rayshawn is applying mulch along the fence in his backyard. For every 3 feet of length along the fence, he needs 2 bags of mulch. The fence is 126 feet long, and he has already finished 18 feet. How many more bags of mulch does he need?

Part A: Number each of the following steps to indicate the order in which they need to be completed to solve this problem.

☐ Divide by 3 to figure our how many more 3-foot segments there are.

☐ Subtract 18 from 126 to find out how many more feet need landscaped.

☐ Multiply by 2 to find the number of bags needed.

Part B: How many more bags of mulch are needed?

- -

5. Samuel went to the movies and purchased a ticket, a bag of popcorn, and a soda. He gave the cashier $20.00 and received $2.56 back in change. Fill in the cost of the popcorn in the table.

Item	Price
Ticket	$9.50
Popcorn	
Soda	$2.62

Countdown: 13 Weeks

1. Adrianna is making a photo collage for her parents' 20th anniversary party. Each poster board can fit 13 pictures, and she has 167 pictures.

   Part A: How many poster boards can Adrianna fill?

   Part B: What is the remainder, and what does it represent?

2. Ahmal is trying to estimate how many boxes he will need to store his miniature car collection in. He has 538 cars, and 27 cars will fit nicely into the boxes he wants to buy.

   Part A: Use the numbers below to choose the best pair that will estimate the number of boxes Ahmal needs, and write them in the blank spaces.

   500   540   530   600   27   30   20

   Part B: Estimate the number of boxes Ahmal will need.

ONLINE TESTING
On the actual test, you might be asked to drag and drop the numbers into the boxes. In this book, you will be asked to write the numbers using a pencil instead.

3. A construction company is looking at a rectangular piece of property on which to build an office building. The area of the property is 20,514 square yards. One side length is 78 yards. Draw the field and label the side lengths.

4. Compare $\frac{8}{10}$ and $\frac{8}{100}$.

   Part A: Shade the decimal to match each fraction.

   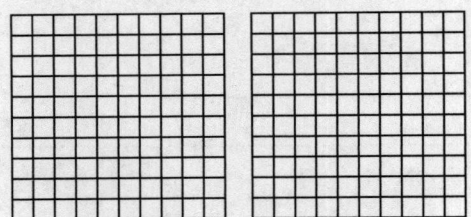

   Part B: Fill in <, >, or =.

   $\frac{8}{10}$ ◯ $\frac{8}{100}$

5. Complete the powers of 10 pattern in the top row of the table below. Then complete the pattern created in the bottom row by writing the corresponding power of 10 with an exponent.

	6,100		610,000	6,100,000
61×10^1	61×10^2	61×10^3		

Countdown: 12 Weeks

NAME ____ **DATE** ____ **SCORE** ____

1. Nick researched the weights of male and female lions. The chart shows his findings. Use rounding to estimate the difference in weight between a male and a female lion.

Male	Female
410.89 lbs	306.21 lbs

2. A coach has timed a swimmer who completed two laps in the pool. The time for the swimmer's first lap was 57.12 seconds. The time for the swimmer's second lap was 61.8 seconds.

Part A: Shade the box under all correct ways of finding the swimmer's total time for both laps.

57.12 + 61.8 ☐   57.12 + 61.80 ☐   57.12 + 61.8 ☐

ONLINE TESTING
On the actual test, you might be asked click in order to shade the boxes. In this book, you will be asked to shade the boxes using a pencil instead.

Part B: Find the total time the swimmer took to swim both laps.

3. Antonio makes $13 for mowing his neighbors' lawns. He is saving up for a telescope that costs $49. Complete the table to help find out how many lawns Antonio will need to mow in order to make enough money to pay for the telescope.

1 Lawn	
2 Lawns	
3 Lawns	
4 Lawns	
5 Lawns	

_____ Lawns

Copyright © McGraw-Hill Education. Permission is granted to reproduce for classroom use.

Grade 5 · Countdown 12 Weeks **29**

4. The chart shows the total number of brownies sold at a bake sale on three different days. Place each of the numbers from the table in the blanks in a way that makes the addition problem the easiest. Explain your reasoning, and find the total.

Day	Brownies Sold
Monday	12
Tuesday	19
Wednesday	8

(____ + ____) + ____

5. Jameson is trying to round 99.9999 to the nearest tenth.

Part A: Jameson asks three friends for the answer and gets three different responses. Circle the correct answer.

100.0

99.0

99.9

Part B: Alana did the same problem but accidentally rounded to the nearest hundredth. She says she got the same answer. Is that possible? Explain.

Countdown: 11 Weeks

1. Write a real-world math problem that can be solved using the base-ten blocks below.

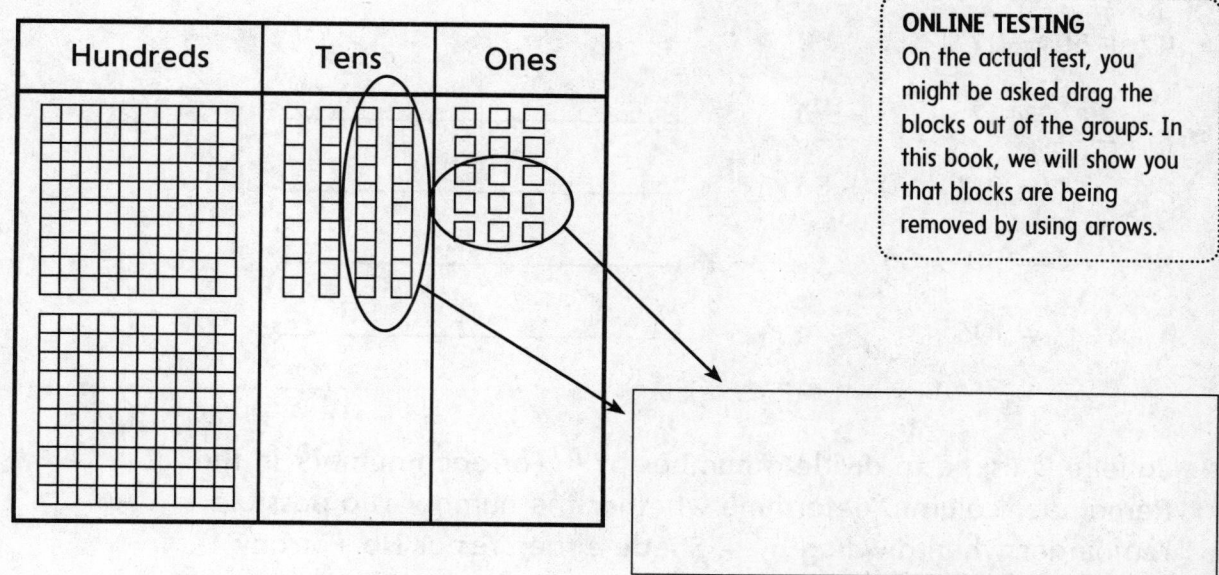

2. Joshua is trying to subtract 8 − 4.13. He sets up the problem like this.

$$\begin{array}{r} 8 \\ -\ 4.13 \\ \hline \end{array}$$

Part A: What is wrong with Joshua's setup?

Part B: What is the answer to Joshua's problem?

3. Look at the solution for doing 47 + 59 mentally. Select from the following properties to fill in the reasons for each step.

Commutative Property	Associative Property	Addition

$$47 + 59 = 47 + (56 + 3)$$

$$= 47 + (3 + 56) \qquad \rule{4cm}{0.4pt}$$

$$= (47 + 3) + 56 \qquad \rule{4cm}{0.4pt}$$

$$= 50 + 56 \qquad \rule{4cm}{0.4pt}$$

$$= 106 \qquad \rule{4cm}{0.4pt}$$

4. Janelle is asked to divide a number by 4. For each number in the Remainder column, determine whether the number is a possible remainder when dividing by 4. Shade either Yes or No. For any number that you marked as Yes, give an example of a division problem that has that number as a remainder when dividing by 4.

Remainder	Yes	No	Example
0	☐	☐	
1	☐	☐	
2	☐	☐	
3	☐	☐	
4	☐	☐	
5	☐	☐	

5. Circle the mistake in the prime factorization tree for 48.

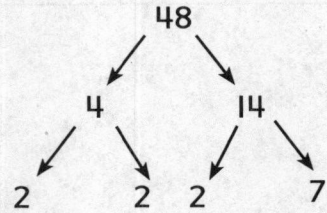

Countdown: 10 Weeks

1. An amusement park costs $47.50 admission for a day. A family of five wants to go to the park. Use the following set of numbers to fill in boxes that will help estimate the total cost for the family.

$40	$30	$50
5	1	10
$150	$250	$350

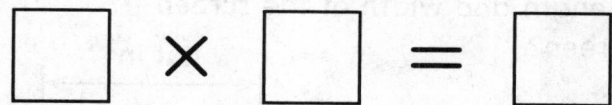

ONLINE TESTING
On the actual test, you might be asked drag the numbers into the boxes. In this book, you will instead write the numbers using a pencil.

2. Shade the models below to calculate 0.8×2.

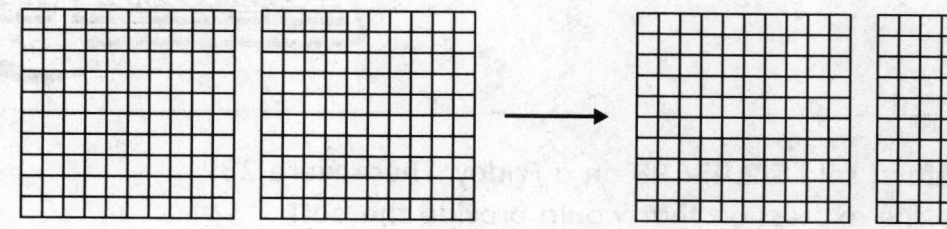

0.8×2 _____

3. Javier sold 9 bags of cookies at $2.25 per bag. Molly sold 5 pieces of pie at $4.15 per piece.

Part A: How much did Javier earn?

Part B: How much did Molly earn?

Part C: Who earned more?

4. Jeremy bought a new computer. The length and width of the screen are shown. What is the area of the screen?

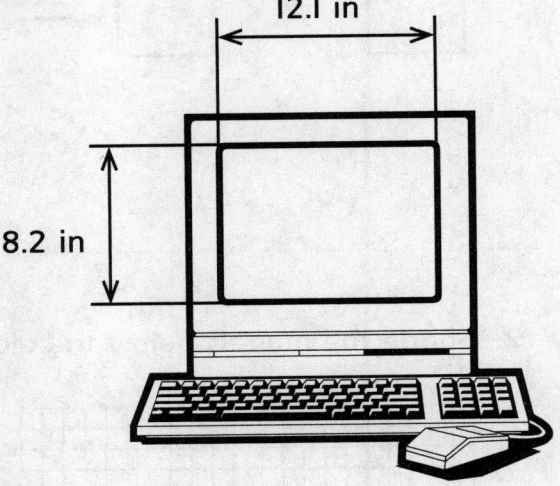

12.1 in

8.2 in

5. A local grocery stand sold $12,456.98 on a Friday. There were 289 customers. Circle the expression that would provide the best estimate for the average amount spent by a customer.

$13,000 ÷ 300

$12,000 ÷ 300

$10,000 ÷ 300

$12,000 ÷ 200

Countdown: 9 Weeks

1. Mr. Jackson took his new car on a family vacation. He drove the car 1,454.625 miles and used 4.5 tanks of gas. How many miles does he get on a single tank of gas?

2. Sort the following multiplication problems into those that have answers that are greater than 1 and those that have answers that are less than 1.

0.89 × 10   0.012 × 10   0.034 × 10²   1.29 × 10³

Greater than 1	Less than 1

ONLINE TESTING
On the actual test, you might be asked drag the numbers into the groups. In this book, you will instead write the numbers using a pencil.

3. Max measured the length of a bug in science class to be 47.61 mm. Write this number in expanded form.

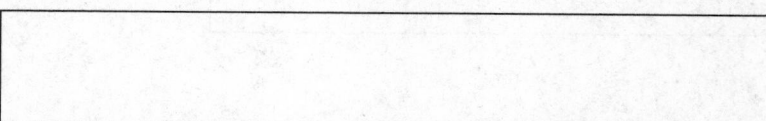

4. Jonathan is trying to calculate $(6.28 \times 50) \times 2$ without a calculator.

Part A: Reorganize the numbers to make the calculation easier.

_____ × (_____ × _____)

Part B: What property did you use to reorganize the numbers?

Part C: What is the answer to Jonathan's question?

....

5. Roland is trying to calculate $4.51 + 12.78$.

Part A: Explain what Roland is doing wrong in his setup?

$$
\begin{array}{r}
12.57 \\
+ \ 4.51 \\
\hline
5.767
\end{array}
$$

Part B: Find the correct solution to Roland's problem.

Countdown: 8 Weeks

1. The bar diagram below can be represented with several expressions. However, not all of the ones below are correct. Circle any that are not correct, and evaluate each expression.

| 6 | 6 | 7 | 7 | 7 | 6 |

$6 + 6 + 7 + 7 + 7 + 6$ $6 \times 3 + 7 \times 3$

_____   _____

$3 \times (6 + 7)$ $3 \times (6 + 7) \times 3$

_____   _____

2. Benjamin wants to find the area of a trapezoid-shaped garden. His teacher told him that the area can be found by first adding the lengths of the top and the bottom, then multiplying the sum by the height, and finally dividing the product by 2.

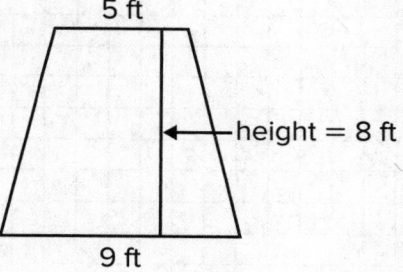

5 ft

←height = 8 ft

9 ft

> **ONLINE TESTING**
> On the actual test, you might be asked shade the boxes by clicking on them. In this book, you will shade the boxes using a pencil instead.

Part A: Shade in the box next to any expression that will find the area of the garden.

☐ $(5 + 9) \times 8 \div 2$ ☐ $[(5 + 9) \times 8] \div 2$

☐ $5 + 9 \times 8 \div 2$ ☐ $5 + 9 \times (8 \div 2)$

Part B: Evaluate the expression to find the area.

3. Both Andrea and Eileen are preparing to run a marathon, a 26.2 mile race. Each of the women begins her training by running 2 miles per day. Andrea says that she will double the amount that she runs per day with each passing week. Eileen says that she will add 5 miles to her daily run with each passing week. Use the two charts to determine who will be running *more than* 26.2 miles per day first.

Andrea	
Week	Miles Per Day
1	
2	
3	
4	
5	
6	

Eileen	
Week	Miles Per Day
1	
2	
3	
4	
5	
6	

4. A farmer is constructing a small fenced in area that can be describe with the ordered pairs (2, 3), (2, 8), (6, 8), and (6, 3). The units for both *x* and *y* are feet. Make a graph. Then find the amount of fencing he will need.

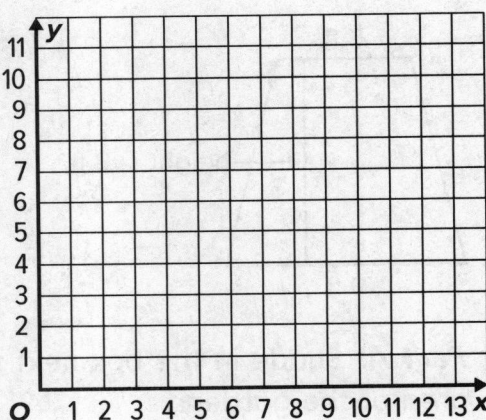

5. A factory that produces piano keys churned out 545,952 keys in 12 months. A piano requires 88 keys. How many pianos can be produced using the keys from the first 3 months?

Countdown: 7 Weeks

1. Anya needs 7 cans of paint to put on three coats in her new living room.

Part A: How many cans of paint will it take to paint a single coat?

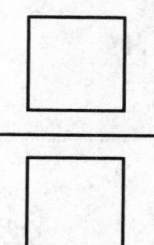

Part B: Place a point on the number line that represents the number.

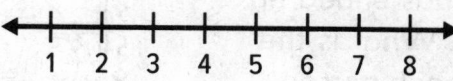

2. 16 fiction books and 20 nonfiction books are to be put in giveaway bags. The number of fiction books in each bag will be the same as the number of nonfiction books in the bag.

Part A: What is the greatest number of bags that can be made?

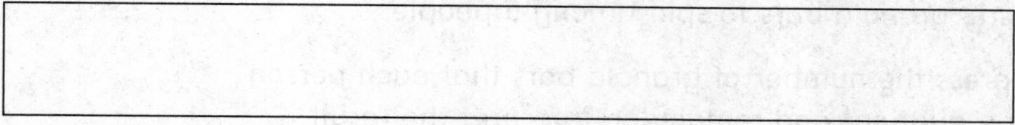

Part B: Jillian says that if the number of fiction books goes up to 18, then the number of bags that can be made will also go up. Is she correct? Why or why not?

3. Mr. McDonald gave his math class the following problem.

   18 pizzas need split among 12 families.

   How many pizzas does each family get?

   Different people in the class gave different answers. Circle the answers that are correct.

   $\frac{18}{12}$ $\frac{9}{6}$ $\frac{1}{2}$ $\frac{3}{2}$ $\frac{9}{4}$ $1\frac{1}{2}$ $1\frac{3}{6}$ $1\frac{1}{3}$

4. A local post office sells stamps in packs of 4, 6, and 7. Andy bought several packs of 4. Erin bought several packs of 6. Jarryn bought several packs of 7. Each of the three friends ended up with the same number of stamps. What is the smallest number of stamps that each person could have purchased?

5. Manuel has 13 granola bars to split among 5 people.

   Part A: Express the number of granola bars that each person receives as a quotient and remainder. Interpret the result.

   Part B: Express the number of granola bars that each person receives as a mixed number. Interpret the results.

Countdown: 6 Weeks

1. The table shows Frances' quiz scores for the past grading period.

   Part A: Fill in the fraction that Frances got correct for each quiz.

Number Incorrect	Number Correct	Fraction Correct
2	6	$\frac{6}{8}$
2	3	
3	7	
2	5	
4	9	

   Part B: Frances' teacher has agreed to drop the worst test score for the grading period. Put her scores in order from least to greatest, and circle the score that can be dropped.

   _____ , _____ , _____ , _____ ,

   > **ONLINE TESTING**
   > On the actual test, you might be asked to drag the numbers into their order. In this book, you will write the number using a pencil.

2. Compare each pair of numbers by using <, >, or =.

   $\frac{14}{20}$ ◯ 0.65 $\frac{6}{25}$ ◯ 0.24 $\frac{1}{50}$ ◯ 0.2

3. Ian and Zion are trying to write $\frac{10}{25}$ as a decimal. Ian says that they should multiply the top and bottom by 4 and then convert to decimal form. Zion says that they should divide the top and bottom by 5, then multiply the top and bottom by 20, and then convert to decimal form. Who is correct? Show the work for each method and give the decimal answer.

4. A runner wants to run 1,000 miles in one year. If he runs the same amount every day, use compatible numbers to estimate the number of miles he should run every day. Show your work.

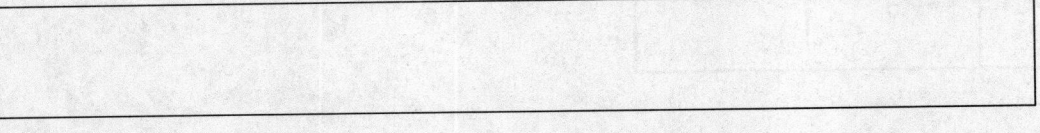

5. Jameson wants to construct a ladder that has 8 rungs. Each rung is 3.2 feet wide. The two sides measure 10.6 feet each. The wood is sold for $2.25 per linear foot. Find how much the wood to construct this ladder will cost Jameson.

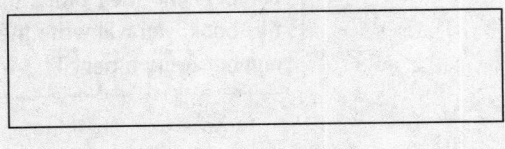

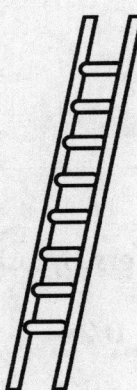

Countdown: 5 Weeks

1. Kalim, Henry, and Joseph agreed to split the lawn mowing for the weekend. Kalim mowed $\frac{5}{12}$ of the lawn. Henry mowed $\frac{5}{12}$ of the lawn. Joseph mowed the rest. Fill in the chart with the fraction of the lawn that Joseph mowed, and put the fraction in lowest terms.

Kalim	$\frac{5}{12}$
Henry	$\frac{5}{12}$
Joseph	

2. Circle the expression that is not equal to the others.

$$\frac{2}{12} + \frac{1}{6} \qquad \frac{2}{12} + \frac{1}{2} + \frac{1}{3} \qquad \frac{1}{6} + \frac{1}{6} \qquad \frac{2}{12} + \frac{1}{12} + \frac{1}{12}$$

3. Theo opened a bag of marbles. $\frac{2}{15}$ of the marbles were red. $\frac{3}{5}$ of the marbles were blue.

Part A: What fraction of the marbles was red or blue?

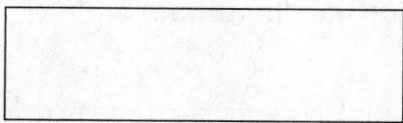

Part B: What fraction of the marbles was neither red nor blue?

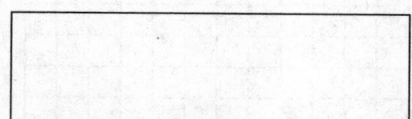

4. Victor claims that if two fractions are in lowest terms, then their sum will be in lowest terms as long as he uses the least common denominator. Drake is sure that he can find two fractions in lowest terms whose sum is not in lowest terms even if he uses the least common denominator.

Part A: Shade in the boxes next to the facts that Drake can use to prove Victor wrong.

☐ $\frac{2}{3} + \frac{1}{2}$

☐ $\frac{1}{4} + \frac{1}{12}$

☐ $\frac{1}{6} + \frac{1}{3}$

☐ $\frac{1}{18} + \frac{1}{3}$

Part B: For each box you shaded, add the fractions together using the least common denominator to show that the sum is not in lowest terms.

5. A company has purchased a large "L" shape plot of land on which to build a new factory. The coordinates of the "L" are (0, 0), (0,9), (4,9), (4,5), (10,5), and (10,0). The units are miles. Plot the "L" shape on the plane, and find the perimeter of the shape.

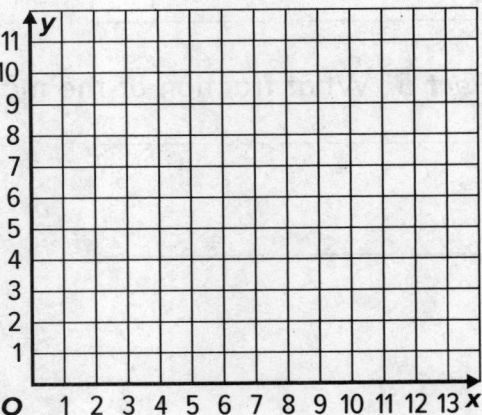

Countdown: 4 Weeks

1. Sydney planted $\frac{5}{9}$ of her fall flowers. She had 63 unplanted flowers to start.

   Part A: How many flowers does she have left to plant?

   Part B: How many more flowers would she need to plant in order to have planted $\frac{2}{3}$ of her flowers?

2. Place each of the following expressions into the two categories of "whole number" and "not a whole number" based on whether or not the product is a whole number.

   $\frac{2}{3} \times 7$ $\qquad$ $\frac{4}{5} \times 15$ $\qquad$ $\frac{6}{7} \times 21$

   $\frac{3}{4} \times 35$ $\qquad$ $\frac{7}{11} \times 100$ $\qquad$ $\frac{9}{13} \times 26$

   ONLINE TESTING
   On the actual test, you might be asked to drag each answer into a bin for each category. In this book, you will write the answer by using a pencil instead.

Whole Number	Not a Whole Number

3. Winston has $\frac{1}{4}$ of a pound of chocolate to split equally among 4 friends.

   Part A: How many pounds will each friend get?

   Part B: How many pounds will two of the friends get together? Write your answer in reduced terms.

4. A playground is to be constructed in the shape of the rectangle shown.

   $\frac{3}{5}$ mile

   $\frac{5}{7}$ mile

   Circle the correct expression for finding the area of the playground. Then find the area.

   $\frac{3}{5} + \frac{5}{7}$ $\qquad$ $\frac{3}{5} - \frac{5}{7}$ $\qquad$ $\frac{3}{5} \times \frac{5}{7}$ $\qquad$ $\frac{3}{5} \div \frac{5}{7}$

   Area = _____ square miles

5. Ms. Trenton measured the rainfall for five consecutive days. Place the days in order from least to greatest amount of rainfall.

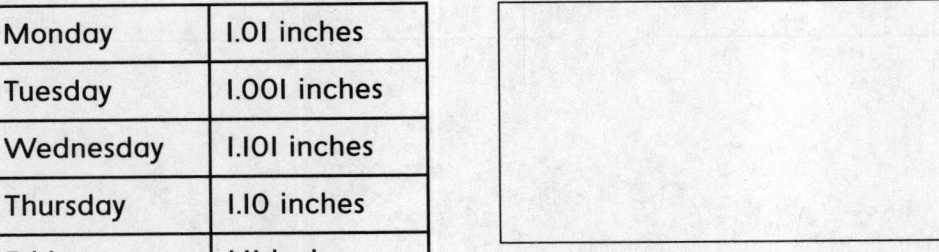

Monday	1.01 inches
Tuesday	1.001 inches
Wednesday	1.101 inches
Thursday	1.10 inches
Friday	1.11 inches

Countdown: 3 Weeks

1. Two teams of scientists measured the length of a cactus needles for a study on desert plant growth. The first team measured the length to the nearest quarter inch and reported a length of $5\frac{3}{4}$ inches. The second team measured the length to the nearest eighth inch and reported a length of $5\frac{7}{8}$ inches. Shade the box under "Yes" or "No" if the length given could be the actual length of the cactus needle.

Yes	No	
☐	☐	5.876 inches
☐	☐	5.85 inches
☐	☐	5.741 inches
☐	☐	5.83 inches

2. In nautical uses, a fathom is a length of measure that is equivalent to 6 feet. Fill in the following conversion chart for fathoms.

_____ fathoms = _____1_____ mile

_____1_____ fathom = _____ yards

_____1_____ fathom = _____ inches

3. Jayne, Carlos, June, and Pedro each measured their dogs' weights, but each used a different measurement. Place the four dogs in order from lightest to heaviest.

Jayne's Dog	42 lbs
Carlos' Dog	624 oz
June's Dog	41 lbs, 7 oz
Pedro's Dog	665 oz

4. A certain species of seaweed doubles in weight every week.

Part A: Fill in the table with the weight for each of the first four weeks. Write your answers as a combination of pounds and ounces.

Week 1	1 pound 5 ounces
Week 2	
Week 3	
Week 4	

Part B: Julian claims that the weight for Week 4 can be written as 10.5 pounds. Is he correct?

5. Bernardo times his drive to and from work. The drive in to work on Monday took 47.23 minutes. The drive home took 56.2 minutes. Circle any correct way of setting up the total time Bernardo spent in the car, and find the answer.

47.23   47.23   47.23   Total = _____ minutes
+ 56.2   + 56.20   + 56.2

NAME ________________ DATE ________

Countdown: 2 Weeks

SCORE ________

1. Fill in <, >, or = to make each of the following statements true.

16 cups ◯ 8 pints

19 quarts ◯ 5 gallons

81 cups ◯ 20 quarts

95 cups ◯ 6 gallons

ONLINE TESTING
On the actual test, you might be asked to drag each symbol into its circle. In this book, you will write the symbol by using a pencil instead.

---

2. Joy started with 2.75 gallons of milk. She used 1.5 pints to make mashed potatoes and another cup to make cookies. How much milks does Joy have left? Give your answers three different ways.

__________ cups

__________ pints

__________ gallons

---

3. There are approximately 3.1 miles in 5 kilometers. Thaddeus is supposed to ride his bike for 20 kilometers for a charity ride.

Part A: How many meters is this ride?

Part B: How many miles is this ride?

4. Place each of the following expressions into the two categories of "Greater than I" and "Less than I" based on the value of the product.

$$\frac{2}{7} \times 4 \qquad \frac{3}{19} \times 5 \qquad \frac{12}{33} \times 3$$

$$\frac{10}{21} \times 2 \qquad \frac{33}{100} \times 3 \qquad \frac{16}{75} \times 5$$

Greater than 1	Less than 1

5. The table below lists the number of pumpkins sold at a pumpkin farm during the course of one week. Estimate how many pumpkins were sold in total for the week. Show how you estimated.

Day	Number of Pumpkins
Monday	212
Tuesday	198
Wednesday	276
Thursday	181
Friday	303
Saturday	315

Countdown: 1 Week

1. One of the figures is an obtuse isosceles triangle. Circle the obtuse isosceles triangle.

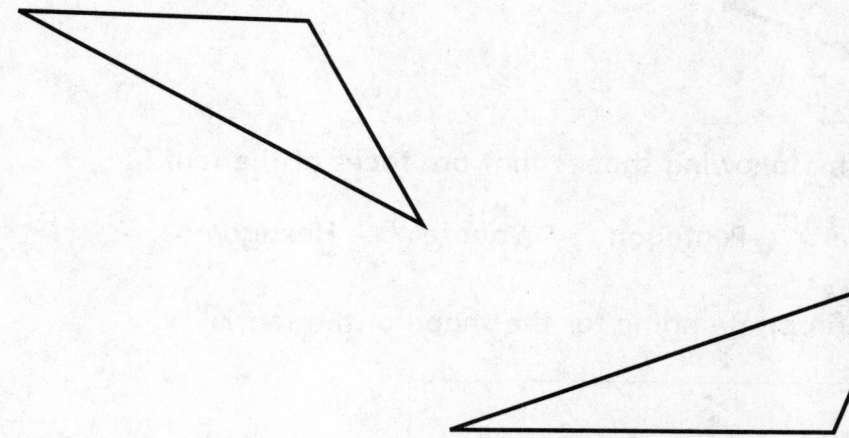

2. An architect is asked to describe the shape of a floor plan for a kitchen, which is shown below.

ONLINE TESTING
On the actual test, you might be asked to drag the names onto the line. In this book, you will write the names by using a pencil instead.

Part A: Write all of the accurate names for the shape of the room in the given space.

Parallelogram	Rectangle
Rhombus	Quadrilateral
Trapezoid	Square

Part B: Which name is the most appropriate for the shape of the room?

3. Shane pitches a tent for his weekend camping trip.

Part A: Circle any of the following shapes that are faces of the tent.

Triangle   Rectangle   Pentagon   Square   Hexagon

Part B: What is an appropriate name for the shape of the tent?

4. Uma is building the following bookshelf. What is the volume of the bookshelf?

$V =$ _____

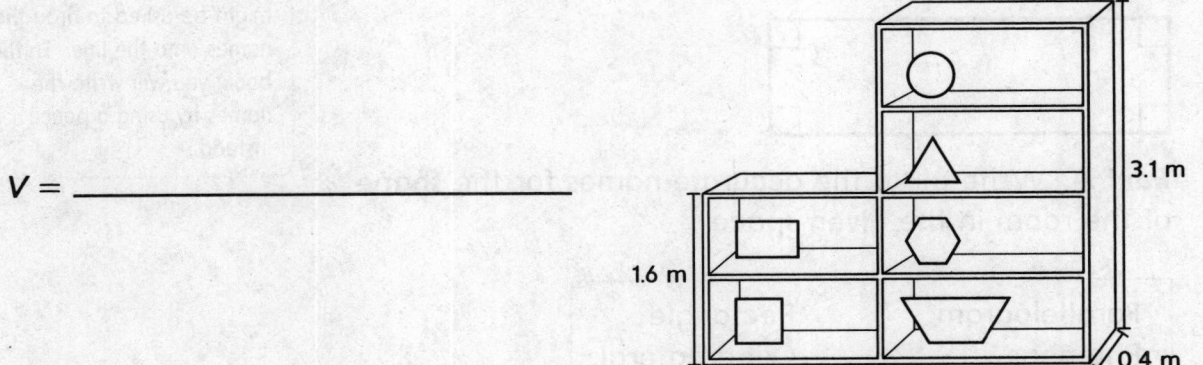

1.2 m

3.1 m

1.6 m

0.4 m

2.3 m

5. A law firm hires the same number of lawyers every year. At the end of 12 years, the firm has hired 48 lawyers. How many lawyers has the firm hired in the last 5 years?

Chapter 1 Test

1. A smart phone company sold 17,468,164 smart phones last year.

   Part A: Fill in the place value chart for the number of smart phones sold by the company.

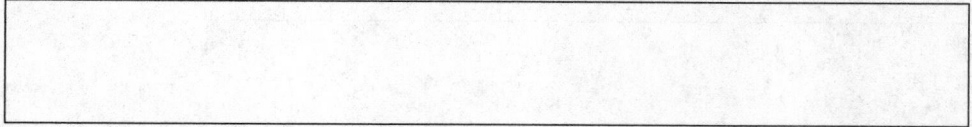

Millions			Thousands			Ones		
hundreds	tens	ones	hundreds	tens	ones	hundreds	tens	ones

10,000,000   7,000,000   400,000   60,000   8,000   100   60   4

   Part B: Write the number in words.

   Part C: Write the expanded form of the number.

2. Alejandro is asked by his teacher to write the smallest five-digit number he can using the digits 1, 3, 5, 7, and 9.

   Part A: If Alejandro is only allowed to use each digit once, what is the smallest five-digit number he can write?

   Part B: If Alejandro is allowed to use each digit more than once, what is the smallest five-digit number he can write?

3. A student measures the length of a postage stamp to be 0.34 inches. He writes down the length as $\frac{34}{1000}$ inches. What is this student's mistake?

4. The table below shows the attendance at a college's first four football games of the season. Put the numbers in order from least to greatest. Is attendance getting smaller or larger?

Date	Attendance
September 28	92,112
September 21	90,912
September 14	88,001
September 7	87,314

5. Shade in the following pictures to show the fractions for 0.3 and 0.30. What can you say about these two numbers by looking at the pictures?

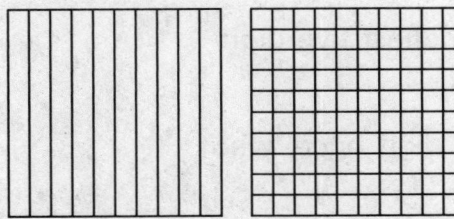

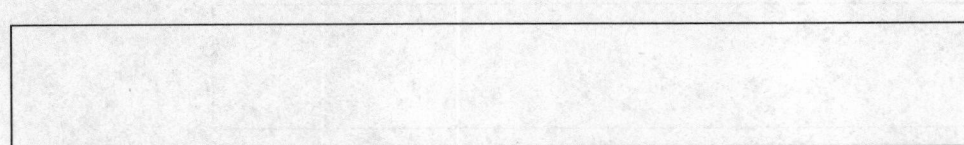

6. Jada's father sent her into the hardware store to find a bolt that is 0.625 inches long. Jada sees the following measurements for bolts. Circle the one is she supposed to buy.

$$\frac{625}{1,000} \qquad \frac{625}{100} \qquad \frac{625}{10,000}$$

7. The table below shows decimals and fractions. Fill in the table so that the left column has equals values as the right column.

0.234	
0.0015	
	$\frac{62}{1,000}$
	$\frac{6}{100}$

8. Paul is weighing a plant for a science project. The weight of the plant is 0.777 kg.

Part A: The value of the digit in the tenths place is how many times the value of the digit in the hundredths place? _____

Part B: The value of the digit in the hundredths place is how many times the value of the digit in the thousandths place? _____

Part C: The value of the digit in the tenths place is how many times the value of the digit in the thousandths place? _____

9. A new player's batting average for the year is 0.289. Write this number out in expanded form.

[ ]

10. Which of the following is not equal to the others? Circle the answer.

4.81

Four and eighty-one hundredths

$4 \times 1 + 8 \times \frac{1}{100} + 1 \times \frac{1}{1,000}$

11. The following chart lists the height of six children from a family. Place the heights in order from greatest to least.

4.25 feet	3.51 feet	3.49 feet
4.2 feet	4.56 feet	3.15 feet

[ ]

12. Sharon writes the weights of her marbles in order from least to greatest, but she makes a mistake. Circle the two numbers that must be switched so that all of the numbers are in the correct order.

1.022 g   1.02 g   1.2 g   1.202 g   1.22 g

13. Mrs. Shen had some eggs in her refrigerator. She bought a pack of twelve eggs for baking. She used six of the eggs and now has nine left in her refrigerator. How many eggs did Mrs. Shen have in her refrigerator before she bought more?

[ ]

14. The Suarez family takes three days to drive to their vacation in North Carolina. The chart shows how many miles the family drove each day. If the family drove 31 less miles on Sunday than they did on Saturday and the total trip was 823 miles, fill in the missing values on the chart.

Day	Miles
Friday	
Saturday	251
Sunday	

15. Janice went out to eat and bought a hamburger, a bag of chips, and a drink. The hamburger cost $2.57, and the chips cost $1.25. Janice gave the cashier $20.00 and received $14.39 in change. How much did the drink cost?

16. A student is struggling to understand the difference between 0.77 and 0.077.

Part A: Explain why 0.77 > 0.077

Part B: Put 0.77, 0.077, and 0.707 in order from least to greatest.

17. *Part A:* Zoe says that "one hundred one thousand" is the same as "one thousand one hundred." Why is she incorrect?

$$\boxed{}$$

Part B: Zoe also thinks that "one hundred one thousandths" is the same as "one hundred and one thousandths." Why is she incorrect?

$$\boxed{}$$

18. Mrs. Hodge has asked her class to use the digits 3, 9, 6, 6, 2, 1 to make a number that is in between 310,000 and 330,000. Four students came up with the following answers. Shade the box next to the answers that are correct.

   ☐ 319,626   ☐ 316,269

   ☐ 321,669   ☐ 328,169

19. Place a decimal point in the following number so that the number is between 34 and 35.

3 4 3 4 3

20. The local news station found out that 123,000 people moved out of the city last year. Shade the box next to the correct way the news reporter should read this number during her report.

   ☐ One hundred twenty-three thousand

   ☐ One hundred and twenty three thousand

   ☐ One hundred twenty three thousandths

   ☐ One hundred and twenty-three thousandths

Chapter 2 Test

1. Jed is buying water bottles for his soccer team. Because all of the packages of water bottles cost about the same price, Jed decides to buy the package of water bottles that provides the greatest total ounces.

   Part A: Complete the table below with the number of ounces per package.

Water Bottle Packages	
Water Bottles in a Package	Ounces per bottle
1	128
12	12
24	8
32	6

Water Bottles in a Package	1	12	24	32
Total Ounces in a Package				

   Part B: Which package provides the greatest total ounces of water? Justify your response.

2. Teams of 4, 5, or 6 members are permitted in a competition. If the grand prize will be divided in whole dollar amounts, evenly among the members of the winning team, which of the grand prizes is possible for this competition?

	Yes	No
$120		
$90		
$48		
$480		

3. Adrianna has 30 bills in her wallet. Some are $1 bills, some are $10 bills, and some are $100 bills. Which of the possible combination of bills in Andrea's wallet has the greatest value? Explain how you solved the problem.

Possible Bill Combinations		
$1 Bills	$10 Bills	$100 Bills
15	15	0
12	17	1
28	0	2
4	26	0

4. The table shows the ticket cost of certain prizes at a fair.

Which combination of prizes can you buy if you earned 432 tickets?

Prizes	Tickets Needed
Stuffed Animals	125
Noise Maker	64
Sticky Hand	38
Pencil	15

Prize Combinations	Yes	No
7 Noise Makers		
3 Stuffed Animals, 2 Pencils		
6 Sticky Hands, 15 Pencils		
2 Stuffed Animal, 4 Sticky Hands		

5. Over the period of one month 159 dogs visited the dog park. Suppose the same number of dogs visited each month for 1 year. How is this total different from the year before when 95 dogs visited the dog park every 3 months for the year? Show your work.

6. *Part A:* Complete the powers of 10 pattern in the top row of the table below. Then complete the pattern created in the bottom row by writing the corresponding power of 10 with an exponent.

780			780,000	7,800,000
	78×10^2	78×10^3		

Part B: Analyze each pattern. Explain the relationship between the top row pattern and the bottom row pattern. What does this pattern mean when considering the numbers above?

7. The following clues are given about a pail of marbles.

1.	There are between 700 and 800 marbles in the pail.
2.	The marbles were purchased in 8 equally-sized bags.
3.	The product of all the digits is 70.

How many marbles are in the pail? Explain how you figured it out.

8. A class will purchase 24 tickets to a play. Each ticket costs $78. Use an area model to find the total cost for the tickets.

Part A: Write an equation to represent the use of partial products to complete each part of the area model.

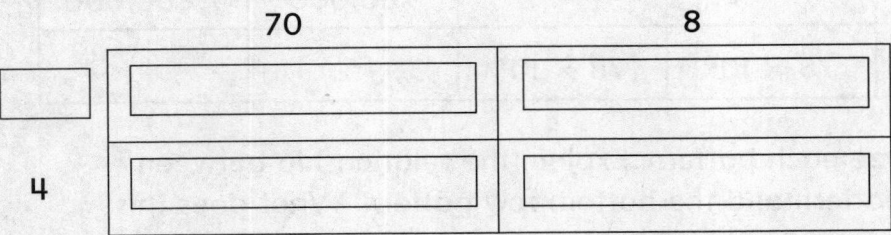

Part B: What is the total cost of the tickets?

9. To the right is an example of Jordan's work on a recent test.

Part A: Identify Jordan's error.

$$
\begin{array}{r}
\overset{\scriptstyle 1\ 6}{1}17 \\
\times\ \ \ 19 \\
\hline
1053 \\
+\ 117 \\
\hline
1170
\end{array}
$$

Part B: Explain how if Jordan estimated the product he would have seen that his answer was not reasonable?

10. A scientist is labeling insects for his collection. He knows the approximate weights of different amounts of each insect. Use the table to complete the weights shown.

10,000 Ants _____ grams

100 Centipedes _____ grams

100 Spiders _____ grams

1000 Honey bees _____ grams

Weights	
1000 Ants	4 grams
1000 Centipedes	140 grams
1 Spider	1 gram
100 Honey bees	1 gram

11. A company makes straws. The table shows the number of straws that are packaged in their different-sized boxes each hour.

    Part A: Complete the table.

    Part B: How would you write the first column of numbers as repeated multiplication expressions?

Number of Straws in Each Box	Number of Boxes	Total Straws
10^2		9,500
10^2	55	
	185	1,850,000
10^3	115	

12. The stairway shown is made by putting 10 cement blocks together. If each cement block costs $23, how much would 10 complete stairways cost? Explain.

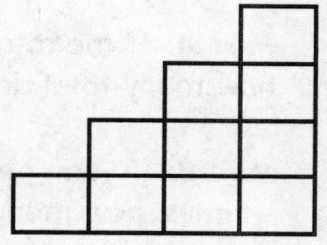

13. A physician recorded a person's resting heart rate to be 87 beats per minute. Complete the table to estimate the total number of times the person's heart would beat for each interval shown.

Number of Minutes	1	10	100	1000
Number of Heartbeats				

14. The product of 54 and another number is 8,720. Use the table to help you estimate the range for the other number.

Multiplication	Product
54 × 10	540
54 × 100	5,400
54 × 150	8,100
54 × 175	9,450
54 × 1,000	54,000

15. Rent costs $478 each month. Complete the partial product diagram for how much rent costs over a year.

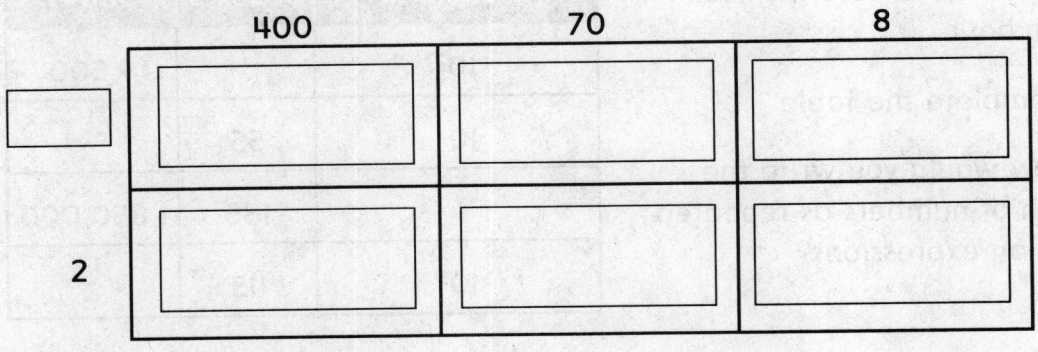

	400	70	8
2			

The rent costs _____ for a year.

16. Skateboarders count rotations in half-turns of 180 degrees.

Part A: If the rotation record is 4 half-turns, how many total degrees is the record?

Part A:

Part B: If Sean performed the rotation record 4 times, how many total degrees did he turn?

Part B:

Part C: Explain how Sean's performance compares to a single half-turn.

Part C:

17. A kilogram is 10^3 grams.

Part A: Write 10^3 grams in expanded notation.

Part B: Suppose a package weighs 2 kilograms. How many grams is it? Explain.

Chapter 3 Test

1. Circle the fact that does not belong to the multiplication fact family.

$$3 \times 9 = 27 \qquad\qquad 27 \div 3 = 9$$

$$3 \times 3 = 9 \qquad\qquad 27 \div 9 = 3$$

2. A group of 36 cans of juice is divided among four children.

   Part A: If each child receives c cans, write an equation to find the unknown.

   Part B: Find the unknown value c.

3. Write and solve a division problem that is modeled by the picture.

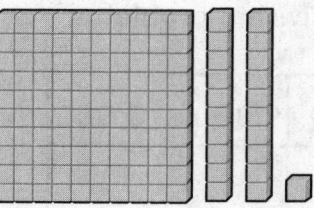 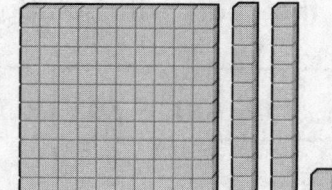

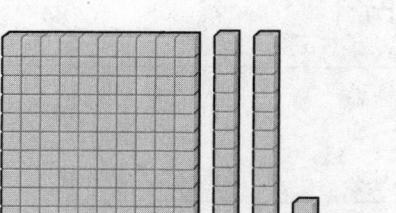

4. Bradford is taking down bulbs from a holiday decoration. The bulbs are put into boxes that can hold 6. He has 81 bulbs. What is the remainder? What is the meaning of the remainder?

5. Circle the mistake in the division problem.

$$
\begin{array}{r}
25R2 \\
4\overline{)92} \\
-8 \\
\hline
22 \\
-20 \\
\hline
2
\end{array}
$$

6. Daiki is trying to sell 40 cupcakes that she made for a bake sale. He would like to sell them in boxes of 6. How many are left over that will need to be sold individually?

7. *Part A:* Fill in the chart with solutions to the division problems.

9,000 ÷ 3	9,000 ÷ 30	9,000 ÷ 300	9,000 ÷ 3,000

Part B: Describe the pattern.

8. Malik says that 16,000 ÷ 4,000 is the same as 160 ÷ 40.
Is Malik correct? Explain.

---

9. Three friends decided to open a household chore business.
They mow lawns, babysit, walk dogs, and clean windows.
The chart shows how much money the business made in the first
month. If the earnings are split equally among the friends, how
much will each person receive?

Mowing Lawns	$67
Babysitting	$82
Walking Dogs	$17
Cleaning Windows	$26

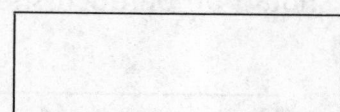

10. Madison and her friend Gabriel are both trying to estimate
182 ÷ 91. They both round to different place values.

Madison	Gabriel
180 ÷ 90	200 ÷ 100

Explain each student's thinking. What are their estimates? Are
both correct?

11. Use base ten blocks to model and solve the division problem
246 ÷ 2.

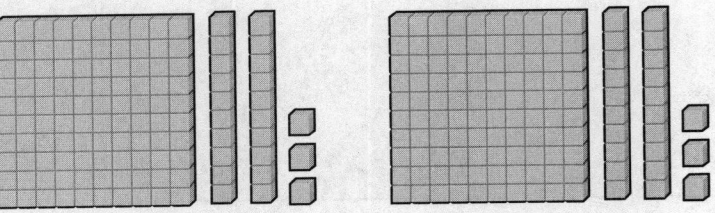

$$246 \div 2 = \underline{\hspace{3cm}}$$

12. A bookshelf has 5 shelves on it. There are 155 books that need
put away.

Part A: Use the distributive property and the picture below to find
how many books should be on each shelf.

5	100	50	5

Part B: Sonny did the problem with a different picture.

	10	10	10	1
5	50	50	50	5

Is he correct? Explain.

13. There are 144 roses that need put onto 8 tables at a wedding reception. How many roses should be put on each table?

14. Nine friends want to go to an amusement park. The total bill for all nine tickets is $423.

    Part A: How much is each ticket?

    Part B: Estimate to check your answer.

15. Daksha was asked by her teacher to predict the number of digits in the following quotients without actually dividing. How can she do this?

    Part A: 834 ÷ 7

    Part B: 567 ÷ 6

16. Describe the student's error in the following division problem, and do the problem correctly.

```
      29
  3)627
   -6
    27
   -27
     0
```


17. Lamar runs 8 miles a day. He wants to know how many miles he ran in a particular month. Is there too much information or not enough information to solve this problem? Shade the box next to the correct description. If there is too much information, name the extra information and solve the problem. If there is not enough information, describe what Lamar would need to know to solve the problem.

☐ Too much information ☐ Not enough information

18. *Part A:* A table at a party seats 8 guests. There are 71 guests expected. Find the number of tables needed and interpret the remainder.

Part B: Christian is making birdhouses. Each birdhouse requires 10 screws. Christian has 81 screws. Find the number of birdhouses Christian can make and interpret the remainder.

Chapter 4 Test

1. Circle any that would not be good ways of estimating 328 ÷ 32.

$$330 \div 30 = 11$$

$$400 \div 20 = 20$$

$$300 \div 30 = 10$$

$$300 \div 20 = 15$$

2. A scientist is studying fish population in an area of the ocean that measures 18,314 square miles. He wants to divide the area into 87 equal size portions to make the study more manageable. Estimate how large each area will be. Show your work.

3. A cab company is interested in how many vehicles it should station outside a particular hotel. This company has vans that hold 14 people. By looking at checkout patterns, the company determined that 160 people leave this hotel per day for the airport. Draw a model with base ten blocks to figure out how many vans the company should have ready in order to seat 160 guests.

_____ vans

4. Mrs. Canzales needs to buy gallons of paint to paint her new house. Each can of paint cost $18. She has $310.

Part A: How many gallons of paint can she buy?

Part B: What is the remainder, and what does it mean?

Part C: Estimate to check your answer. Show your work.

5. Quentin is asked by his teacher to write a division problem with quotient of 23 and a remainder of 3. Quentin wrote the following problem: $347 \div 15$.

Part A: Complete Quentin's problem to show that it is not correct.

Part B: Help Quentin fix his problem by changing the 347 by a small amount.

6. A factory produces 12,376 granola bars in 52 minutes. How many granola bars does the factory produce per minute?

7. A school recently received a donation for $7,072. The school has 17 different student organizations and wants to split the gift evenly. Miles has been asked to help figure out how much each organization should get. He starts the problem off like this:

$$
\begin{array}{r}
3 \\
17\overline{)7,072} \\
-5\,1 \\
\hline
1\,9
\end{array}
$$

At this point, Miles knows that the 3 is not correct because 19 is bigger than 17. He adjusts the 3 to a 2 and gets

$$
\begin{array}{r}
2 \\
17\overline{)7,072} \\
-3\,4 \\
\hline
3\,6
\end{array}
$$

Miles is now confused. What did he do wrong?

[ ]

8. Carlos' goal is to keep track of the total amount that he has run. After 16 months, his total is 2,400 miles. His coach, however, wants to know how much he ran in the past year. If Carlos ran the same amount every month, find his total distance for the past year.

[ ]

9. **Part A:** Fill in the following table with quotients and remainders.

Division Problem	Quotient	Remainder
6245 ÷ 12		
6246 ÷ 12		
6247 ÷ 12		

Part B: What pattern do you notice?

[ ]

10. The college marching band raised $15,708 to help pay for a trip to a national parade. There are 132 students in the band. How much will each student receive in order to help pay for his or her airfare? Will there be any left over?

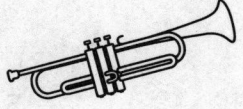

11. An art gallery purchases the same amount of prints per year to sell in its gift shop. In the last 7 years, the gallery has purchased a total of 882 prints. How many did the gallery purchase in its first 3 years?

12. A swimming pool is being designed that is 22 feet wide and 42 feet long. The shallow end will be 22 feet wide and 28 feet long. What will be the area of the deep end?

13. Tyron is asked to find the missing value h in the equation:
$12,336 \div h = 16$
His friend Seamus says that he can rewrite this using another member of the same fact family and then solve the problem.

Part A: Fill in the boxes to rewrite the equation using another member of the fact family.

$$\boxed{} \div \boxed{} = \boxed{}$$

Part B: Find the missing value h.

$h = \boxed{}$

14. A restaurant sells chicken in packs of 6 pieces. The restaurant orders a large bag of chicken and splits the pieces into packs of 6, but there are some pieces left over. Shade the boxes next to any number that is a possible remainder, then explain your reasoning.

☐ 0   ☐ 3   ☐ 6

☐ 1   ☐ 4   ☐ 7

☐ 2   ☐ 5   ☐ 8

15. ABC Electronics produces a circuit board that can be used in computers. ABC's factory produced 18,270 boards last week. ABC supplies these boards to 90 different computer manufacturers and wants to give an equal amount to each manufacturer.

Part A: Fill in the division fact with compatible numbers to estimate how many boards each manufacturer should receive.

☐ ÷ ☐ = ☐ boards

Part B: Find the exact number of boards each manufacturer will receive.

Part C: Is your estimate greater than or less than the actual number? Explain how you could have known this ahead of time.

16. 2,134 ÷ 8 has a remainder of 6. Circle all of the following facts that also have remainders of 6.

2,126 ÷ 8   2,127 ÷ 8   2,128 ÷ 8

2,142 ÷ 8   2,143 ÷ 8   2,144 ÷ 8

17. Hyun usually runs races that are 10 kilometers long. His time for a 10 kilometer race is 50 minutes. There is a charity race this weekend that is 9 kilometers long. What should Hyun expect his time to be at the race this weekend?

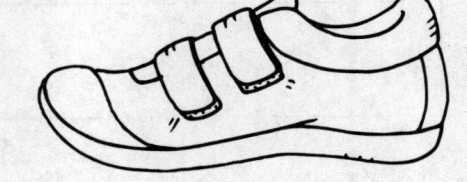

18. There is a set of swings to split among 17 different playgrounds in a city. Each playground gets 4 swings, and there are 3 left over.

Part A: How many swings are there in all?

Part B: Write a division problem to model this situation.

19. Write *always, sometimes,* or *never* for each of the following statements.

The remainder is less than the divisor. _____

The quotient is greater than the remainder. _____

The divisor is equal to the dividend. _____

20. A charity has raised $14,569 for use in food banks around the country. There are 17 different food banks that will split the funds. How much money does each food bank receive?

NAME .... DATE ....

SCORE ....

Chapter 5 Test

1. Write a number that rounds to 2.3 when rounded to the nearest hundredth *and* when rounded to the nearest tenth, but does not round to 2.3 when rounded to the nearest thousandth.

2. Garret's teacher asked him to round the following number to the nearest tenth. Write Garret's answer in both expanded form and standard form.

$$3 \times 10 + 4 \times 1 + 2 \times \frac{1}{10} + 4 \times \frac{1}{100}$$

Expanded Form

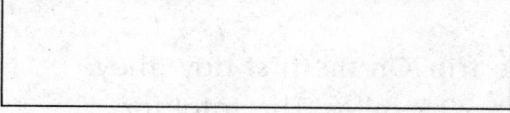

Standard Form

3. Maria claims that it does not matter whether she first rounds two numbers to the nearest hundredth and then adds them or whether she first adds the two numbers and then rounds to the nearest hundredth. Is she correct? Explain.

4. Veronica is buying the following items from the grocery store.

Milk	$2.60
Bread	$1.80
Crackers	$3.51
Cheese	$6.78

Part A: Round each term to the nearest dollar to estimate the total bill.

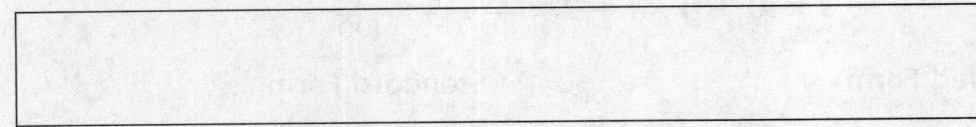

Part B: Is the estimate greater than or less than the exact total? How do you know?

5. Ophelia and her father took a three-day bike trip. On the first day, they rode 27.5 miles. On the second day they rode 23.2 miles. The total for the trip was 65.7 miles. Ophelia wants to know about how long the ride was on the third day. Is this a question about an exact answer or an estimate? Answer Ophelia's question.

6. Joaquin used 10-by-10 grids to model 1.79 + 1.36. He is stuck. Describe what Joaquin must do next, then find the sum.

   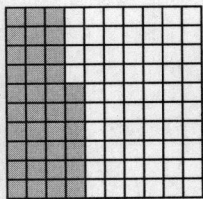

7. Look at the following model. Circle the expressions that could match the model.

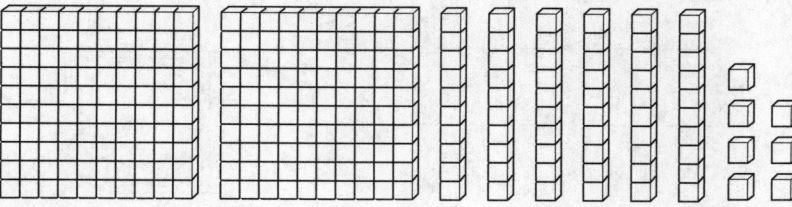

1.27 + 1.4   2.52 + 0.15   1.6 + 1.7

1.66 + 1.1   1 + 1.67   1.07 + 1.60

8. Tax on two purchases was $0.36 and $0.87.

Part A: Shade the model to show the regrouping needed to find the sum of the two values.

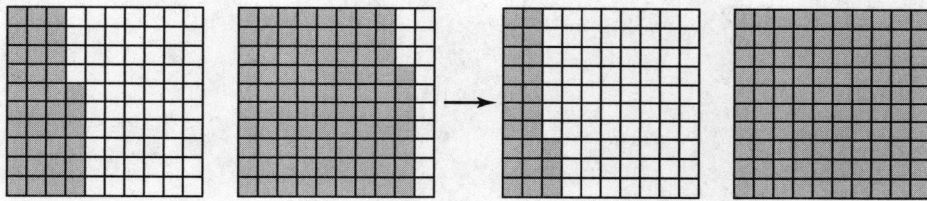

Part B: What is the sum?

9. On a class trip to Washington D.C. the bus made four stops for gas and spent the following amounts. What was the total gas bill for the trip?

| $123.57 |
| $135.88 |
| $132.19 |
| $98.27 |

10. Kiara bought two items from a music store. The first item was between $8.00 and $9.00. The total of the items was $20.21. Write down two different possibilities for the cost of each item.

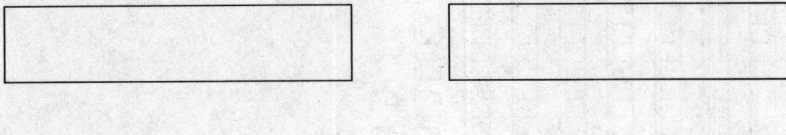

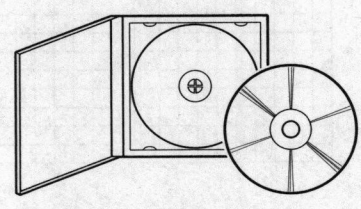

_____

11. Shade the box under "Yes" or "No" to indicate whether each problem will require regrouping.

Yes	No	
☐	☐	3.61 + 0.71
☐	☐	41.23 + 21.32
☐	☐	33.51 + 33.5
☐	☐	4.09 + 12.5

12. Write the correct property of addition for each step.

$9.9 + (3.6 + 4.1) + 0$

$= 9.9 + (4.1 + 3.6) + 0$ _____

$= (9.9 + 4.1) + 3.6 + 0$ _____

$= 14 + 3.6 + 0$ Addition

$= 17.6 + 0$ Addition

$= 17.6$ _____

13. Circle the problems that will require regrouping.

13.71 − 2.8	3.4 − 2.2	145.65 − 140.05
65.67 − 13.91	245.16 − 5.16	123.45 − 11.111

14. The local college baseball stadium recorded the number of people in attendance at the first three games of the season.

Game I	14,998 people
Game 2	10,672 people
Game 3	15,002 people

Part A: Write down the best order in which to add the numbers so that it is easiest to find the total using mental math.

Part B: Find the total attendance for the first three games.

15. Write and solve a word problem for the following place-value chart.

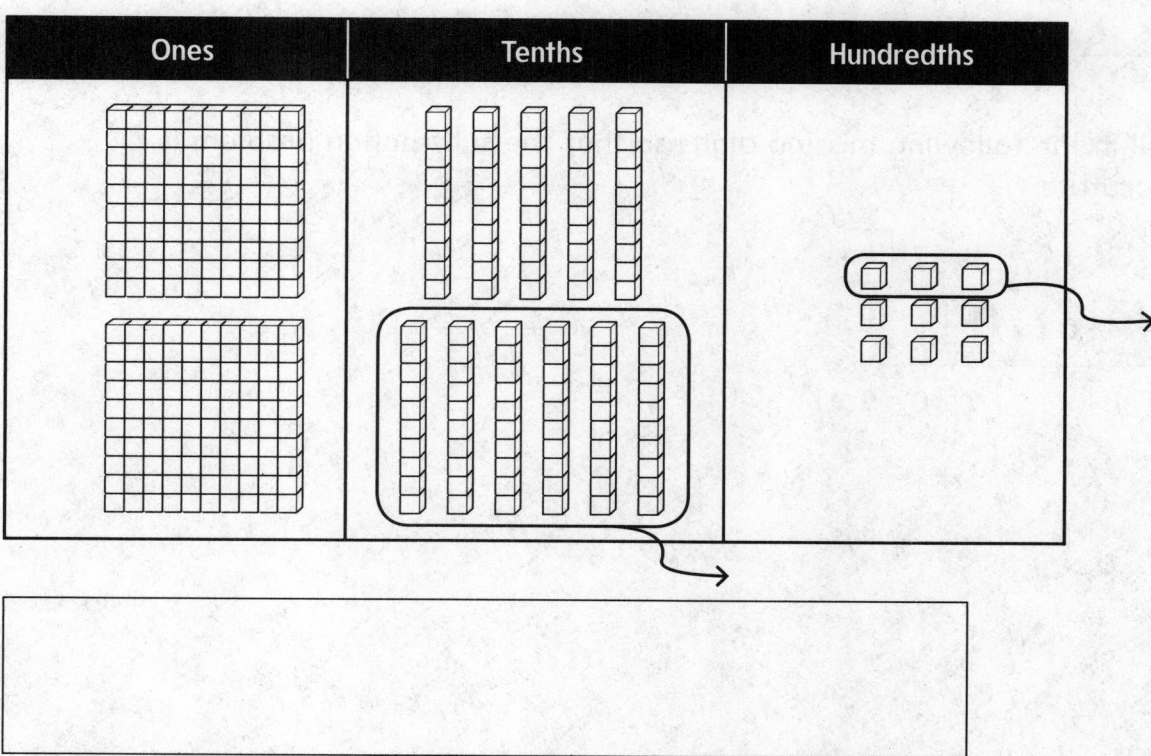

Ones	Tenths	Hundredths

16. A woodworker has purchased 92 linear meters of wood for framing. Each frame takes 18.21 linear meters of wood to make. Fill in the following chart to determine how many frames the woodworker can make with 92 linear meters of wood and how much wood will be left over.

Frames Made	Wood left over
1	92 − 18.21 = 73.79 meters
2	
3	
4	
5	

_____ frames

_____ meters left over

17. A baker needs to have 15.5 pounds of flour on hand for the weekend bread sale. He looks in the cupboard and finds a bag that has 4.6 pounds, a bag that has 6.1 pounds, and a bag that has 4.7 pounds. Does the baker have enough? If so, how much extra does he have? If not, how much more does he need?

18. Fill in the following missing digits so that the subtraction problem is accurate.

```
   3  1  .  4  2 ☐
−  ☐  0  .  ☐  1  7
─────────────────
   1  ☐  .  7  0  9
```

Chapter 6 Test

1. The number of people who bought admission to the community pool last Friday was 198. Admission was $4.95. Estimate the amount of money that the pool brought in.

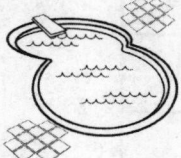

2. Tamar walks 0.3 miles to school every day. Regroup and shade the models to figure out how far he walks in five days.

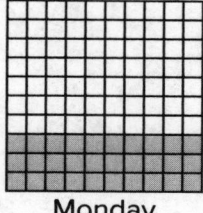    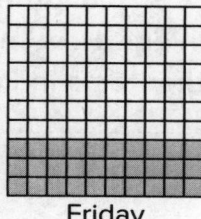

Monday   Tuesday   Wednesday   Thursday   Friday

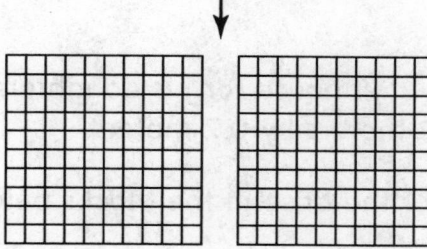

_____ miles

3. Sort each problem into the box that lists the correct number of decimal places in the answer.

34.15×12 4.67×11 $4.56 \times 12,345$

0.45×23 4×2.5 45×123.4

0 Decimal Places	1 Decimal Place	2 Decimal Places

4. Bihn is selling the following items at a garage sale.

Part A: How much will he make if he sells 5 CDs, one bike, 15 books, and 2 pairs of skates?

CDs	$1.25
Bike	$25.16
Books	$0.76
Skates	$9.12

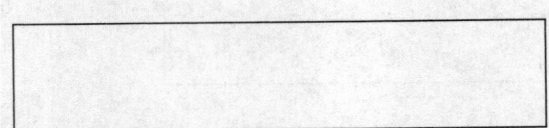

Part B: Bihn intends to purchase a $65 video game with his earnings. How much more money does he need if he sells all of the items in Part A?

5. Mr. Olson is building a new doll house for his daughters. The base of the doll house will be 1.8 meters by 0.7 meters.

Part A: Shade the region of the base in the blocks below.

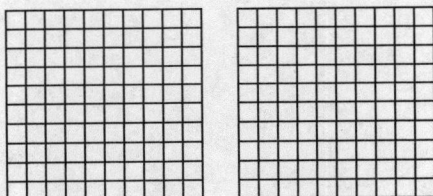

Part B: Rearrange the shading to help figure out the area of the base.

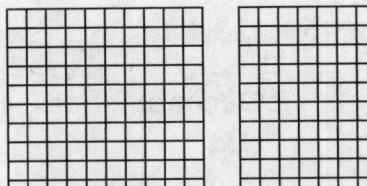

_____ square meters

6. A runner can run a mile in 6.27 minutes. If the runner maintains this pace, how many minutes will it take the runner to run 26.2 miles?

7. The amount of sugar in a serving of each of three brands of cookies is shown in the table. Which contains more grams of sugar: 1 serving of Brand A, 1.5 servings of Brand B, or 2 servings of Brand C?

Brand	Grams of Sugar per Serving
A	33.13
B	22.6
C	15.6

8. Malik is tiling his kitchen floor. The kitchen measures 13.25 feet by 11.75 feet.

Part A: What is the area that needs tiled?

Part B: Tile comes in packs that will cover 10 square feet. How many packs will Malik need to buy?

9. Put the following numbers in order from least to greatest.

1.23×10^3 123.0×10^2 0.0123×10^4

10. Yasmine reads 12 pages of her book the first day, 18 pages the second day, 24 pages the third day, and so on. If the pattern continues, on what day will Jasmine finish her 264 page book?

[box]

11. Harris is asked by his teacher to multiply the following in his head.

$(2.5 \times 7) \times (2 \times 4) \times 50$

Part A: Rewrite the expression so that the multiplication is easier to do in your head.

(_____ × _____) × (_____ × _____) × _____

Part B: Find the answer.

[box]

12. Light bulbs come in a pack of 6 that costs $12.29. Estimate the price per light bulb.

[box]

13. Santiago was given the following diagram that is supposed to represent a decimal division. Complete the division equation.

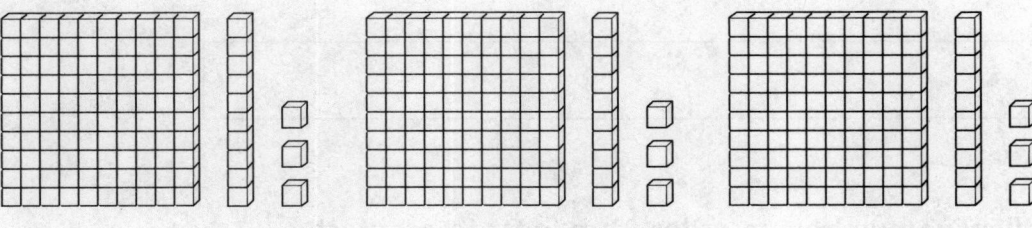

_____ ÷ _____ = _____

14. Hot dogs are sold in packs of various sizes. Which of the three brands is the best buy?

Brand	Number in Pack	Price
A	8	$4.48
B	10	$5.56
C	12	$6.12

15. A jeweler purchased 1.8 feet of gold chain to make bracelets. If each bracelet requires 0.6 feet, draw a model to help find how many bracelets he can make.

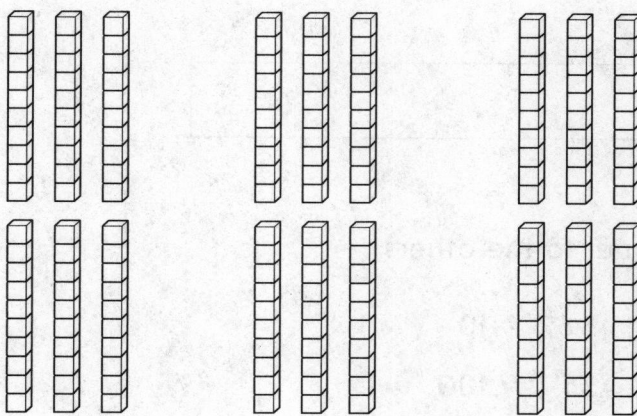

_____ bracelets

16. The area of a picture frame is 16.625 square feet. The length is 3.5 feet. Find the width.

17. Daniel claims that a decimal divided by a decimal can never be a whole number. Is he correct? If so, explain why. If not, give an example showing that he is wrong.

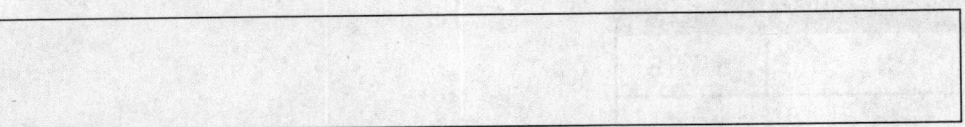

18. On Monday, a driver purchased 18 gallons of gas for $61.38.

Part A: What was the price of gas per gallon that day?

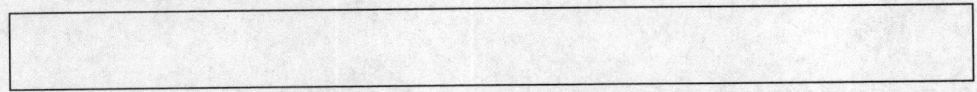

Part B: The next day, the price of gas was $3.29 per gallon. How much could the drive have saved had he waited until Tuesday to buy gas?

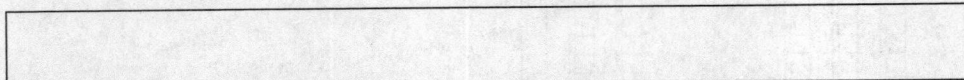

19. Circle the expression that is not equal to the others.

567 ÷ 1,000   5.67 ÷ 10

56.7 ÷ 10,000   56.7 ÷ 100

20. In July of 2014, Phoenix, Arizona had a record rainfall of 5.04 inches in 12 hours. On average, how many inches of rain fell per hour?

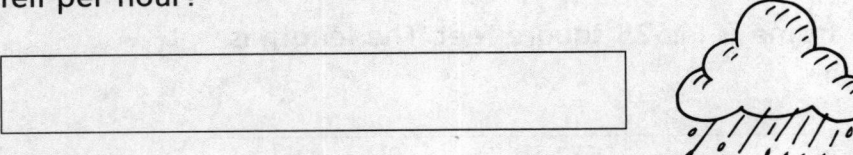

Chapter 7 Test

1. Gary watches cars go by his house. He counted 4 red cars, 3 blue cars, 4 yellow cars, 4 white cars, 3 black cars, and 4 green cars.

   Part A: Write and evaluate an expression for the total number of cars Gary saw using only addition.

   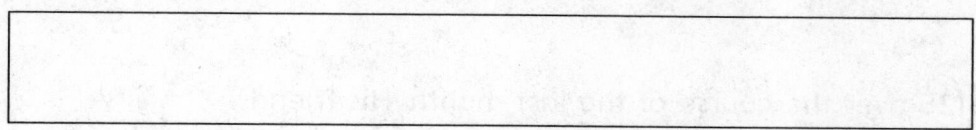

   Part B: Write and evaluate an expression for the total number of cars Gary saw using both addition and multiplication.

   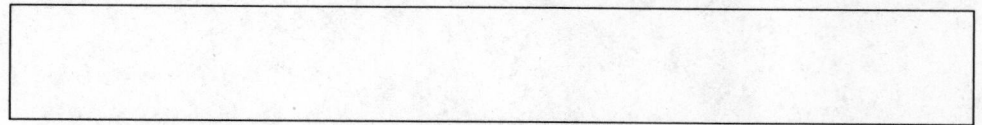

2. A football is thrown up in the air. The height of the football after two seconds is $3 \times 2^2 + 4 \times 2 + 6$. Find the height of the football.

   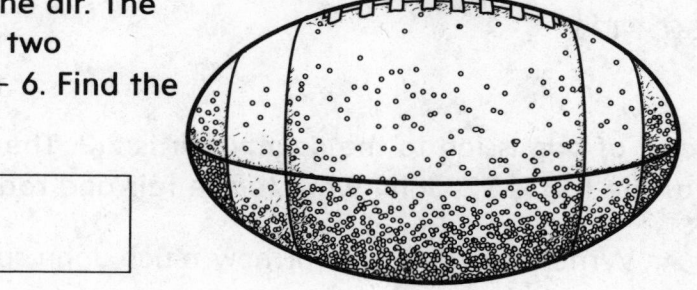

3. Sammy's teacher asked him to evaluate the expression $2 + 4 \times 5$. Sammy wrote down the answer of 30. What did Sammy do wrong? Give the correct answer.

   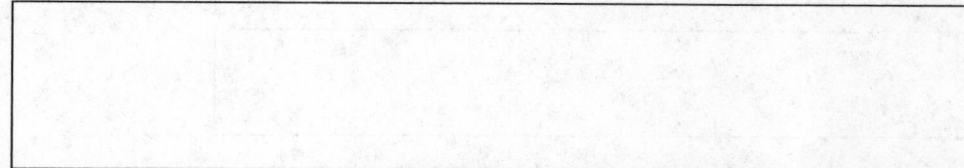

4. Camilla and her three friends bought three tickets to the movie theater at $8 each. They also bought a large popcorn for $10. They split the bill evenly. Write and evaluate an expression for the total amount that each friend spent.

5. Jared saved $125 over the course of the last month. His friend Hector saved twice the difference between Jared's amount and $50. Circle which of the following expressions represents the amount that Hector saved in the last month.

$2 \times 125 - 50$

$(50 - 2) \times 125$

$2 \times (125 - 50)$

$2 \times 50 - 125$

6. The cost of admission to the county fair is $12. The cost of each ride at the fair is $2. John went to the fair and rode 14 rides.

Part A: Write an expression for how much John spent.

Part B: Evaluate the expression to find out how much John spent.

7. A pizza shop sells a single pizza for $11. However, at the end of the night, they will sell the extra pizzas for $6 each. If the pizza shop sold 51 pizzas before the discounted price and made $615, find the number of discounted pizzas they sold.

8. Joline decided to sell her china doll collection. She sold the dolls for $24 each. Joline sold 8 dolls to Sylvia, and then sold half of the remaining dolls to Jane. She made $288.

Part A: How many dolls did Joline have before she decided to sell her collection?

Part B: How many dolls does Joline have now?

9. In which pattern will the numbers go over 100 first? Write out the patterns to show your answer.

Pattern A: Start at 2 and multiply by 2.

Pattern B: Start at 55 and add 10.

10. Look at the number of blocks in the following pattern.

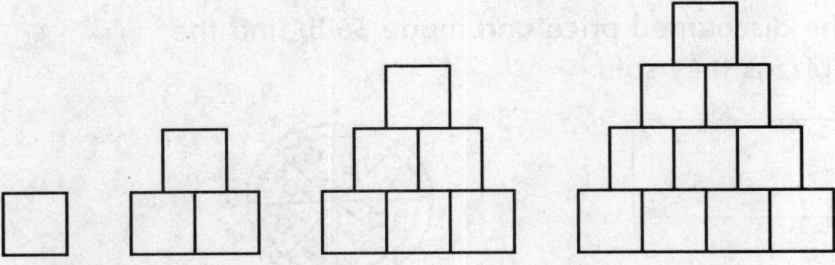

Part A: Write down the pattern for the number of blocks in each stack.

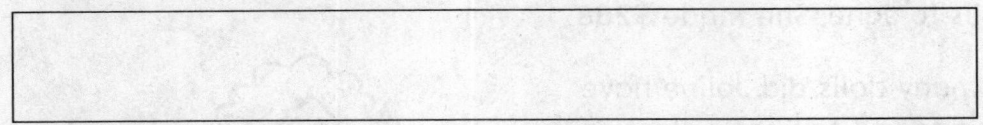

Part B: Predict how many blocks will be in the next stack.

11. Circle the pattern that does not belong.

3, 9, 27, 81   5, 15, 45, 135

2, 5, 8, 11   2, 6, 18, 54

12. Mrs. Gerard saves $10 per week. Mr. Gerard saves $8 per week. Fill in the following table for the total amount in savings at the end of each week, and describe the pattern for the total savings.

Week	Mrs. Gerard	Mr. Gerard	Total Savings
1	$10	$8	
2	$20	$16	
3			
4			

Pattern: _____

13. A park ranger is taking a tour of the major sites in a national forest to check for safety violations. The map below shows the sites.

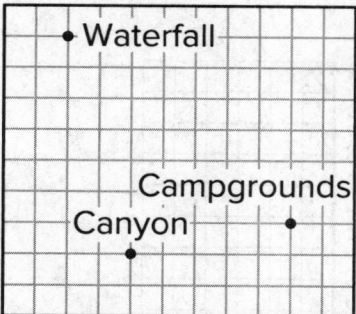

Part A: If the ranger starts at the canyon, describe the path he can take first to the waterfall, and then to the campgrounds if he must follow the grid lines.

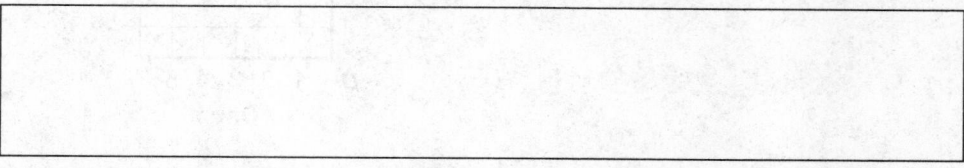

Part B: If the ranger wants to go directly from the canyon to the campgrounds, how many units shorter is that than going to the waterfall first? Again, the ranger must follow the grid lines.

14. Brock is trying to make a plan for building a table. He wants the top to be a rectangle. He has placed three of the corners on the coordinate plane below.

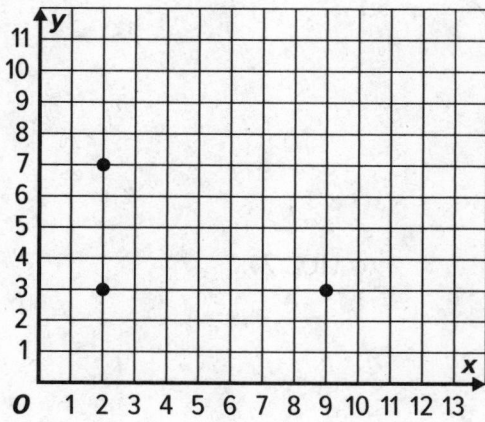

What is the coordinate of the fourth corner? _____

15. Jackie and Taylor decide to start exercise routines. Jackie does 5 situps on the first day and adds 2 each day. Taylor does 3 situps on the first day and adds three each day.

Part A: Find the number of situps that each girl does for the first four days, and graph them as ordered pairs.

Part B: On which day does each girl perform the same number of situps?

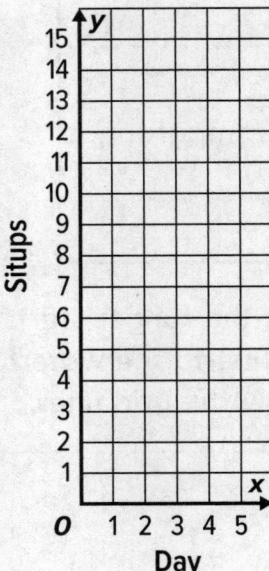

Situps

0 1 2 3 4 5

Day

16. Mr. Zhao walks from the grocery store to his house in a straight line. The grocery store is at point (1,2). His house is at point (11,7). For the ordered pairs listed, circle all of the order pairs that Mr. Zhao passes through on his way home.

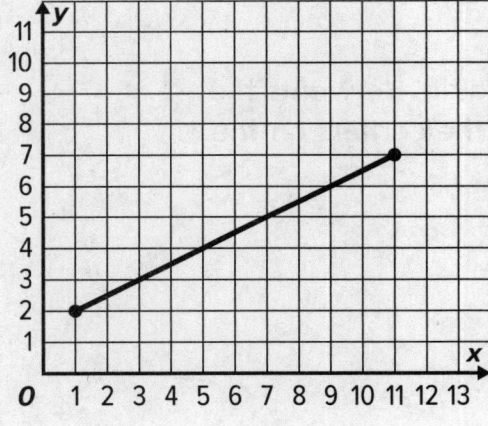

(3, 3)   (4, 5)   (7, 5)   (9, 6)   (10, 7)

Chapter 8 Test

1. Mr. King used 16 gallons of gas in 5 days. Circle any of the following that describe the average amount of gas Mr. King used per day.

   $\dfrac{16}{5}$ $\dfrac{5}{16}$ $3\dfrac{1}{5}$ $5\dfrac{1}{3}$

2. A carpenter is framing 12 windows in a house. He used 51 feet of wood on the project. If the windows are all the same size, how many feet of wood were used in a single window? Write the answer in two different ways using fractions.

3. Consider the following five numbers.

   12   15   36   18   8

   Part A: What is the greatest common factor of all five numbers?

   Part B: Cross out two of the five numbers so that the remaining numbers have a greatest common factor of 6.

4. The table below shows the number of flowers that a florist has to put in vases for a wedding display. Each vase will have only one color of flower, and the florist wants to make sure that each vase has the same number of flowers. If the florist puts all of the flowers in vases, what is the greatest number of flowers that could be in each vase?

Pink	18
Red	30
White	24

5. Deandre has made a table of his baseball card collection based on the year. Fill in the third column with the fraction of his total cards that year represents, and put the fractions in lowest terms.

Year	Number of Cards	Fraction of Total Cards
2009	12	
2010	18	
2011	10	
2012	20	

6. Shade the box next to any fraction that is in simplest form.

☐ $\frac{8}{12}$ ☐ $\frac{9}{12}$

☐ $\frac{10}{12}$ ☐ $\frac{11}{12}$

7. Dai went shopping and received $1.55 in change in quarters and dimes. He told his friend Ginny that the change came in 11 coins and asked her to guess how many of each coin he had. Ginny guessed that there were 5 quarters and 3 dimes.

Part A: Is Ginny correct? Explain

8. Mrs. Franklin took her five children to an amusement park. The cost of tickets for the 2 younger children was $12.50 each. The cost of her ticket was $15.00. She spent a total of $80.50. What was the cost of each ticket for her three older children?

Part B: If Ginny's guess is not correct, find the correct answer.

9. Isabelle buys gas every five days. If she buys gas today and today is a Saturday, how many more days will it be before she buys gas on a Saturday again?

10. Mr. Sanchez goes to the movies every 18 days. His brother goes to the movies every 30 days. If they were at the movies together this evening, how many more days will it be before they are at the movies together again?

11. Students from three different fifth grade math classes were asked to return permission slips for a field trip. The table shows what fraction of each class turned in their permissions slips on the first day. Number the classes in order from 1 to 3 starting with the class that had the smallest fraction of students turning in the slips.

Mrs. Haley's class	$\frac{5}{12}$
Mr. Black's class	$\frac{5}{11}$
Mrs. Mayne's class	$\frac{2}{5}$

_____ Mrs. Haley's class

_____ Mr. Black's class

_____ Mrs. Mayne's class

12. The local football coach has always had a goal of winning $\frac{3}{4}$ of the games in a season. The numbers below show the fraction of games won for the season. Circle those seasons below when the coach met his goal.

2011–2012 Season
$\frac{8}{11}$

2012–2013 Season
$\frac{10}{13}$

2013–2014 Season
$\frac{10}{14}$

13. A baseball player got 7 hits in his last 25 at bats. Shade the model to help write the decimal that represents the fraction of at bats resulting in a hit.

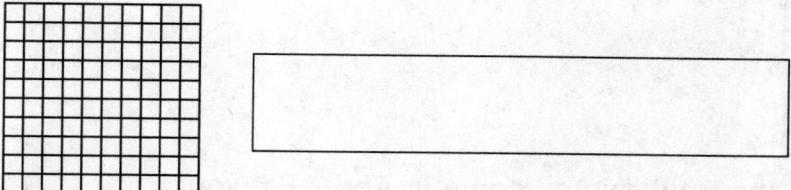

14. Circle all of the numbers that could be modeled with the following block.

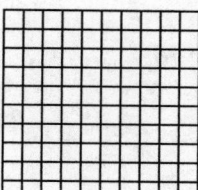

$\dfrac{50}{100}$ $\dfrac{10}{20}$

$\dfrac{17}{35}$ 0.5

15. Match each fraction with its decimal equivalent.

$\dfrac{13}{25}$ 0.25

$\dfrac{2}{5}$ 0.4

$\dfrac{7}{20}$ 0.52

$\dfrac{1}{4}$ 0.35

16. Felipe built a model train that has a scale of $\dfrac{21}{25}$. Write this number as a decimal.

17. The table shows the amount of rainfall that a town experienced in the last three months.

April	10 inches
May	9 inches
June	6 inches

Part A: What fraction of the total rainfall came in April? Put your answer in lowest terms.

Part B: What fraction of the total rainfall came in May? Write your answer as a decimal.

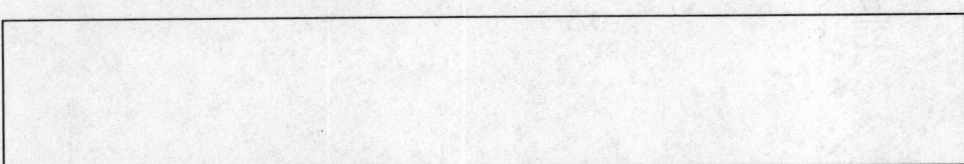

18. Jaylon is asked to find the greatest common factor of three numbers. He finds the greatest common factor of the first two numbers to be I. Jaylon claims he knows the answer without any extra work. How does he know?

NAME   DATE   PERIOD

Chapter 9 Test   SCORE

1. Frederick has been asked to sort five bags into three categories based on whether they are closest to "empty", "half full", or "full". Each fraction represents how full the bag is. Write each of the fractions in the correct box.

$$\frac{2}{19} \qquad \frac{2}{5} \qquad \frac{2}{7} \qquad \frac{6}{11} \qquad \frac{10}{12}$$

About Empty	About Half Full	About Full

2. The following chart shows how much it snowed in the first three hours of a snowstorm. Find the total snowfall in inches for the first three hours. Write your answer in lowest terms.

First Hour	$\frac{1}{8}$ inch
Second Hour	$\frac{3}{8}$ inch
Third Hour	$\frac{2}{8}$ inch

3. Tamika bought red beans, pinto beans, and black beans. She purchased $\frac{15}{16}$ pounds of beans in total. If Tamika bought $\frac{3}{16}$ pounds of red beans and $\frac{5}{16}$ pounds of pinto beans, find the weight of black beans she purchased. Write your answer in lowest terms.

4. Keisha bought $\frac{1}{6}$ tank of gas in the morning on her way to work. She added $\frac{3}{4}$ tank on her way home from work. Fill in the model below to determine what fraction of a tank she bought altogether.

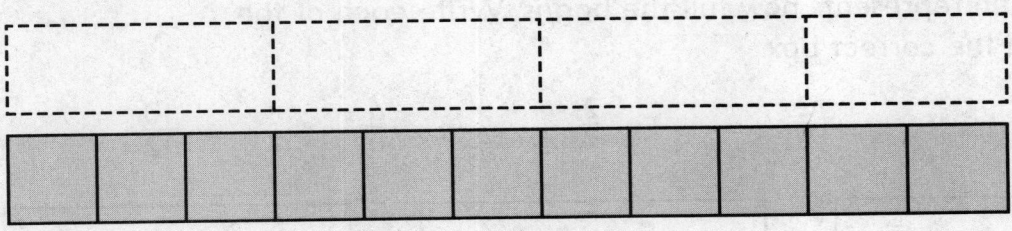

_____ tank

5. Feng was asked by his teacher to add $\frac{1}{5}$ and $\frac{3}{10}$. Feng got an answer of $\frac{6}{5}$.

Part A: Compare the two fractions to $\frac{1}{2}$ and show that Feng cannot be correct.

Part B: What is the correct answer in lowest terms?

6. Circle which of the following does not belong, and describe why.

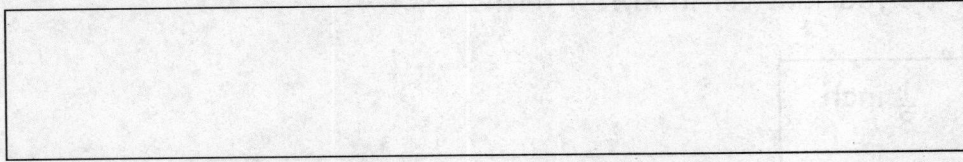

$\frac{1}{3} + \frac{1}{2}$ $\left(\frac{1}{6} + \frac{1}{6}\right) + \frac{1}{2}$ $\frac{1}{3} + \left(\frac{1}{6} + \frac{1}{6} + \frac{1}{6}\right)$ $\left(\frac{1}{2} + \frac{1}{2} + \frac{1}{2}\right) + \frac{1}{2}$

7. Mrs. Prim had $\frac{3}{5}$ of a cup of flour. She used $\frac{1}{2}$ cup on a dessert recipe.

   Part A: Fill in the model below to help find how much flour Mrs. Prim has left.

   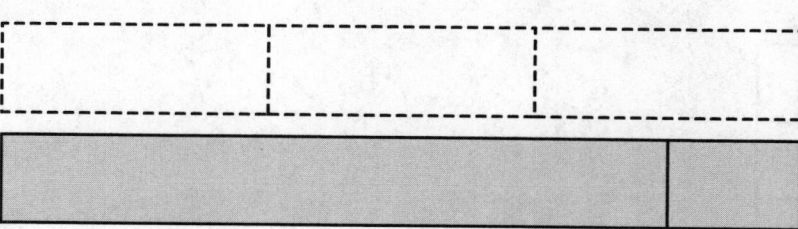

   _____ cup

   Part B: Mrs. Prim needs another $\frac{1}{4}$ cup for a sauce recipe for the dessert. Does she have enough flour left? Explain.

8. Luciano walked $\frac{7}{18}$ mile on Saturday morning. On Sunday, she walked $\frac{5}{9}$ mile. How much more did she walk on Sunday than on Saturday? Shade the box next to any correct answer.

   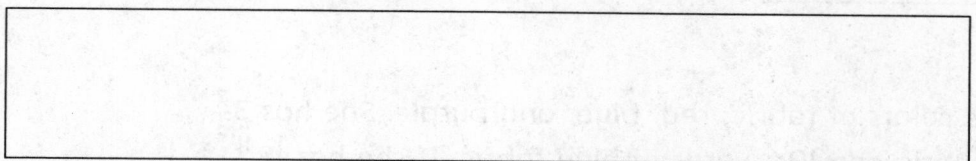

   ☐ $\frac{3}{18}$ ☐ $\frac{1}{6}$ ☐ $\frac{2}{12}$

9. Fill in the box to make a true number sentence.

$$\frac{11}{12} - \frac{\Box}{3} = \frac{1}{4}$$

10. A field goal kicker practices by moving back the same number of yards every time he kicks. On the third kick he is 40 yards away from the goal posts. On the sixth kick he is 49 yards away. Fill in the following chart and work backwards to figure out how far away the kicker was on the first kick. Circle the answer.

Kick #1	
Kick #2	
Kick #3	40 yards
Kick #4	
Kick #5	
Kick #6	49 yards

11. Mia has three colors of fabric: red, blue, and purple. She has $3\frac{7}{8}$ yards of red fabric and $10\frac{2}{15}$ yards of blue fabric. If she has $19\frac{11}{13}$ yards of fabric al together, about how much purple fabric does she have? Show how she estimated.

12. Write and solve a problem involving addition and mixed numbers. Put all mixed numbers in lowest terms.

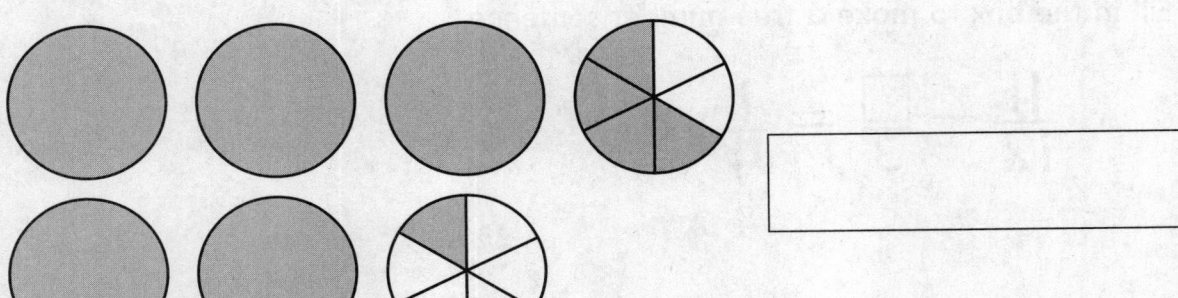

13. Jasmine is making a fruit smoothie that requires the following amounts of certain juices.

Apple Juice	$3\frac{1}{4}$ cups
Raspberry Juice	$\frac{3}{8}$ cup
Grape Juice	$\frac{1}{2}$ cup

Part A: Find the total amount of juice that this recipe will make.

Part B: If Jasmine already has $1\frac{1}{2}$ cups of apple juice, how much more will she need to buy in order to have enough for the recipe?

14. A construction crew is painting lines on the side of a new highway. In one week they are supposed to have $67\frac{1}{2}$ miles completed. On Monday, they painted $13\frac{5}{6}$ miles, and on Tuesday they painted $12\frac{1}{3}$ miles. How many more miles are left to paint?

15. Circle which problems would require renaming.

$$4\frac{1}{3} - 2\frac{1}{2} \qquad 5\frac{5}{6} - 3\frac{1}{3} \qquad 12\frac{3}{8} - 7\frac{2}{5}$$

16. Julian has three lengths of rope: $12\frac{1}{8}$ feet, $13\frac{3}{8}$ feet, and $11\frac{7}{8}$ feet.

   Part A: Estimate the total length of rope that Julian has by rounding to the nearest whole foot.

   Part B: Estimate the total length of rope by rounding to the nearest half foot.

   Part C: Find the exact length of rope that Julian has in total.

17. In the following model fill in the tiles with fractions that make the quantities equal.

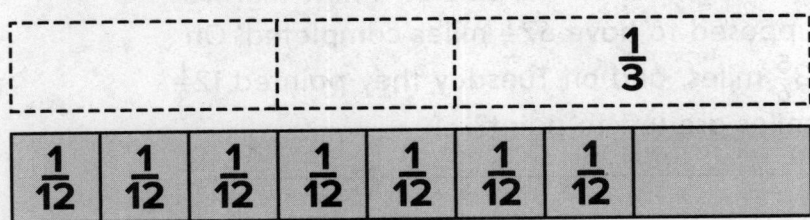

18. Rashaun purchased a 5 gallon bucket of paint. He spilled some paint when opening the bucket. There are $3\frac{3}{8}$ gallons left in the bucket. How much paint did Rashaun spill?

Chapter 10 Test

1. Santana took out 24 books from the library. She returned $\frac{3}{8}$ of them.

   Part A: Draw a model to illustrate the situation.

   Part B: How many books does Santana have left?

2. Yolanda says that $\frac{3}{5}$ of 2 is the same as $\frac{2}{5}$ of 3. Draw bar models to show that she is correct.

3. A cook has $11\frac{7}{8}$ cups of flour. He uses $\frac{2}{3}$ of his flour on Saturday morning. Estimate how many cups he used. Show how you estimated.

4. Look at the following model.

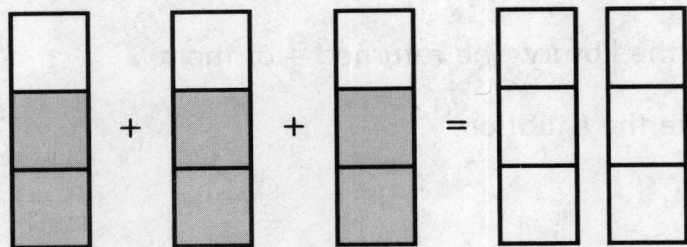

Part A: Shade the blocks on the right side of the equal sign so that the model represents a true statement.

Part B: Write a multiplication problem and answer for the model.

5. A carpenter bought 11 linear feet of an oak board. He used $\frac{2}{5}$ of the board on a baseboard. How many feet did he use?

6. Thelma has 8 yards of fabric. She used $\frac{2}{3}$ of the fabric on a project. Circle any answer that shows how many yards Thelma has left.

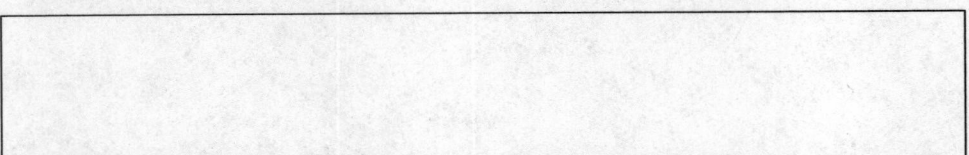

$\frac{16}{3}$ yards $5\frac{1}{3}$ yards $\frac{8}{3}$ yards $2\frac{2}{3}$ yards

7. A field is $\frac{2}{3}$ miles long and $\frac{1}{2}$ mile wide. Draw a model to help find the area of the field.

Area = _____

8. An electrician has $\frac{14}{15}$ yard of wire. He used $\frac{5}{7}$ of the wire.

Part A: How many yards of wire did the electrician use?

Part B: How many yards of wire does the electrician have left?

9. The area of a triangle can be found by multiplying $\frac{1}{2}$ times the base times the height. Find the area of the triangle.

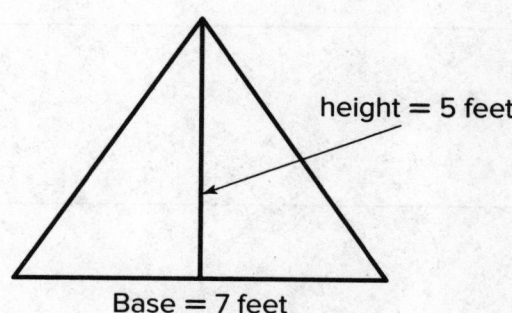

height = 5 feet

Base = 7 feet

Area = _____

10. Jentilly bought $3\frac{2}{3}$ pounds of black cherries. She bought $1\frac{1}{2}$ times that amount in red cherries.

    Part A: How many pounds of red cherries did Jentilly buy?

    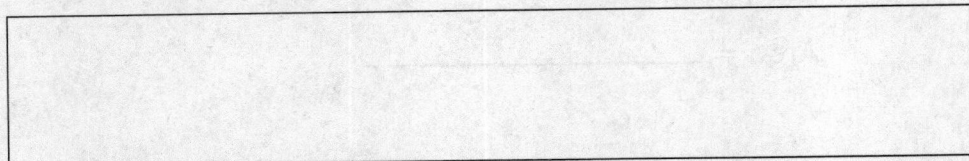

    Part B: How many pounds of cherries did Jentilly buy in total?

    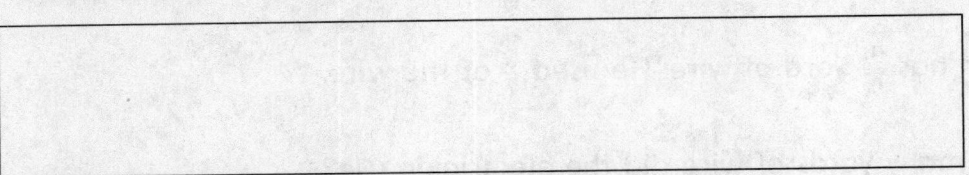

11. Fill in the circles with <, >, or =.

    $10 \times \frac{2}{3}$ ◯ 10 $17 \times \frac{7}{4}$ ◯ 17

    $12 \times 1\frac{2}{5}$ ◯ 12 $8 \times \frac{9}{9}$ ◯ 8

12. Kaitlin spent $\frac{2}{3}$ of an hour on homework. Her sister, Judy, spent $1\frac{1}{2}$ times that amount. Explain why Judy's homework time is between $\frac{2}{3}$ of an hour and $\frac{1}{2}$ hour.

13. Chase needs to cut his fishing line into pieces that are $\frac{1}{4}$ of a foot. He has 3 feet of fishing line. Use fraction tiles to help figure out how many pieces Chase can make.

_____ pieces

14. Mrs. Chen slices 5 pies. Each slice represents $\frac{1}{8}$ of a pie. How many slices of pie are there?

15. Tyrone has $\frac{1}{6}$ pound of almonds. He wants to split this equally among 5 people. What fraction of a pound will each person get?

16. Gretchen claims that $\frac{1}{3} \div 4$ is the same as 4 divided by $\frac{1}{3}$. Is she correct? Explain.

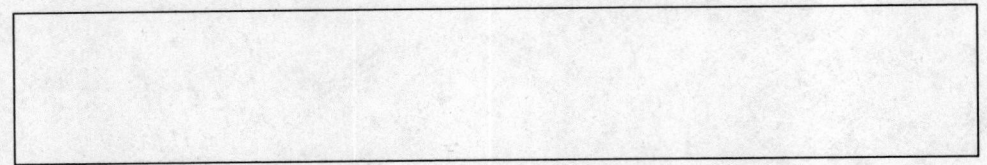

17. Ishmael bought a brand new bag of green, yellow, and red marbles. There are 40 marbles in the bag. $\frac{1}{4}$ of the marbles are yellow. The number of green marbles is the same as the number of red marbles. How many of each color are there?

_____ red marbles

_____ yellow marbles

_____ green marbles

18. Sort each of the fraction problems into those for which the product or quotient is greater than I, those for which the product or quotient is less than I, and those for which the product or quotient is equal to I.

$$\frac{1}{4} \div 3 \qquad\qquad 2 \div \frac{1}{5} \qquad\qquad 2\frac{1}{5} \times 3$$

$$\frac{1}{7} \times 7 \qquad\qquad \frac{1}{8} \times 2 \qquad\qquad 3\frac{1}{6} \times \frac{1}{2}$$

Greater than I	Less than I	Equal to I

Chapter 11 Test

1. Chelsea is measuring a piece of thread.

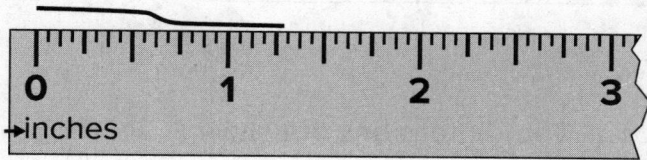

   Part A: What is the length measured to the nearest half inch?

   _____

   Part B: What is the length measured to the nearest quarter inch?

   _____

2. A biologist determined the wingspan of a baby bird to be $6\frac{3}{4}$ inches when measured to the nearest quarter inch. Her lab partner measured the same wingspan to the nearest half inch and got 7 inches. Circle the statement that describes the actual wingspan.

   Less than $6\frac{3}{4}$ inches   Between $6\frac{3}{4}$ and 7 inches

   Greater than 7 inches

3. Compare using <, >, or =.

   144 in. ◯ 12 ft   6.5 yd ◯ 18 ft

   2 mi ◯ 10,561 ft   72 in. ◯ 2 yd

4. A runner ran a marathon, which is 26.2 miles. How many feet is this?

```

```

5. Mr. Schnur is carpeting his living room. The dimensions are show in the diagram. However, the carpet company wants to know how many square inches this is. Find the area of the floor in square inches.

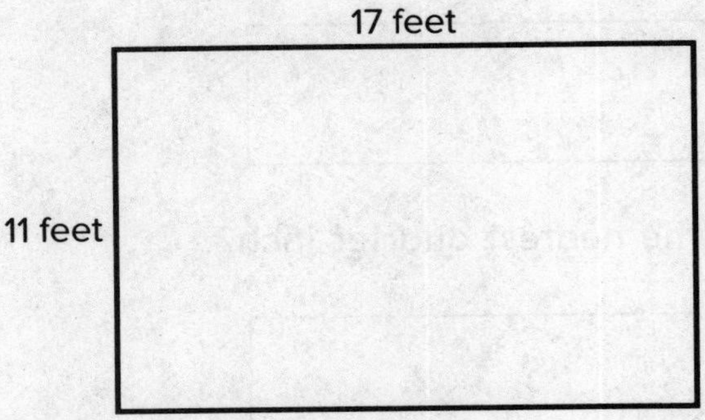

17 feet

11 feet

```

```

6. Francis measured the weight of a red block and got 5 ounces. He then found that a green block weighs the same as 4 red blocks.

Part A: What is the weight of the green block in ounces?

```

```

Part B: What is the weight of the green block in pounds and ounces?

```

```

Part C: What is the weight of the green block in pounds?

```

```

7. How much heavier is 7 pounds than 110 ounces?

8. Frederico has three bags of flour with the following weights.

Bag 1	15 ounces
Bag 2	14 ounces
Bag 3	11 ounces

How many pounds of flour does Frederico have in all?

9. A car manufacturer makes two models. Model A weighs 1.5 T. Model B weighs 2,900 pounds.

Part A: Which model weighs more?

Part B: What is the difference between the two weights in pounds?

10. Circle the statements that are false.

7 pints > 14 cups   2 gallons > 17 pints

9 cups > 2 quarts   1 gallon = 16 cups

11. Calista is making tea for a tea party. Each serving will be 1 cup. How many pints of tea will she need for 16 guests?

12. Dr. Blanchard recommends drinking a gallon of water a day. His patient, Sam, has an 8 fluid ounce glass. How many glasses does Sam need to drink a day in order to follow the doctor's recommendation?

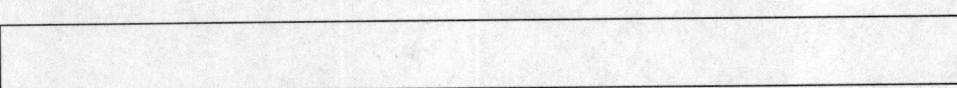

13. One gallon of a particular liquid weighs $8\frac{1}{4}$ pounds.

Part A: How many ounces does one gallon of the liquid weigh?

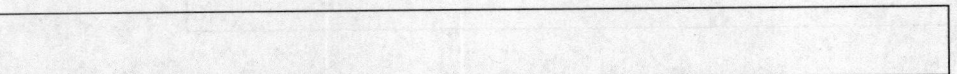

Part B: How many ounces does one quart weigh?

Part C: How many ounces does one pint weigh?

14. The line plot shows the snowfall in inches for the last fourteen days in the town of Snowshoe.

Part A: Make a new line plot that shows the snow fall in feet.

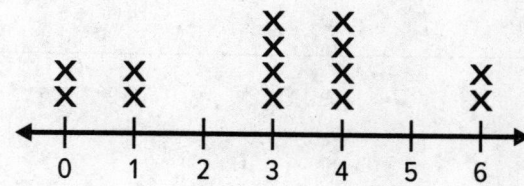

Part B: What is the fair share in feet if the same amount of snow fell every day?

15. Match the item with the appropriate unit of measurement.

The length of a race   millimeter

The width of a postage stamp   centimeter

The thickness of a penny   meter

The length of a table   kilometer

16. A small table for a dollhouse requires 6 cm of a thin wooden board. A dollhouse maker has 12 meters of wood. How many tables can he make?

17. A large rectangular field has length 1.5 km and width 0.75 km. What is the area of the field in square meters?

18. Shanice weighed her collection of books as 12,014 grams. Her brother, Marquis weighed his books and got 13 kg. Who has the greater weight in books? By how much?

19. If a farmer has 386 grams of tomatoes, 671 grams of potatoes, 711 grams of zucchini, and 997 grams of yellow squash, find the total weight in vegetables in kilograms.

20. Circle the quantity that is not equal to the others.

7.1 L   7,100 mL   7 L 100 mL   1 L 700 mL

Chapter 12 Test

1. What shapes make up the surface of a soccer ball?

2. A regular triangle has one side length of 15 cm. Find the perimeter of the triangle.

3. A farmer is making a triangular corn maze for a fall attraction. Fill in the boxes next to the words that describe the triangle.

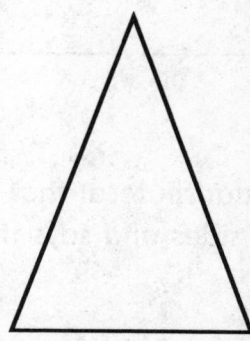

☐ acute   ☐ obtuse   ☐ right

☐ scalene   ☐ isosceles   ☐ equilateral

4. A construction company has marked off a site for a building in the shape of a quadrilateral. All four sides of the site are congruent. Opposite angles are congruent, but there are no right angles. Draw a possible shape for this site.

5. For each of the following pairs of quadrilaterals, describe one thing they have in common.

   A rhombus and a square _____

   A trapezoid and a rectangle _____

   A square and a rectangle _____

   A parallelogram and a
   rhombus _____

6. Jorge says he is thinking of a quadrilateral that has two right angles and at least one set of parallel sides and says that it must be a rectangle.

   Part A: Draw a picture that proves Jorge wrong.

   Part B: Name your shape.

7. A box company is experimenting with nets for making their boxes. Circle any of the following that will fold up to make a box.

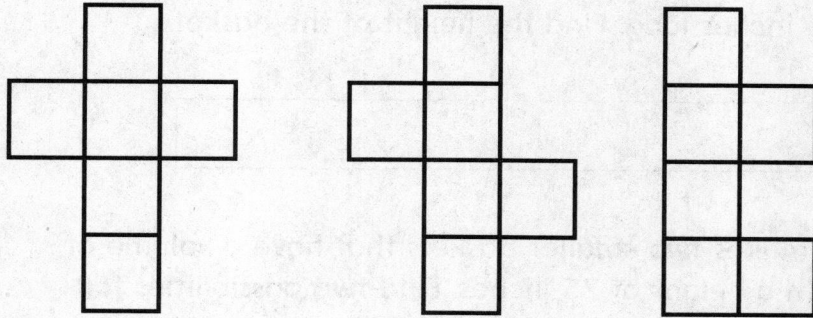

8. A bricklayer is laying bricks for a new patio.

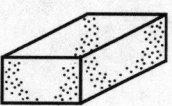

Part A: Fill in the information below for the brick.

_____ faces _____ edges _____ vertices

Part B: What is the name of the shape?

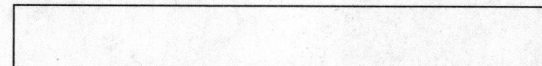

9. Each cube has a side length of 1 inch. Find the volume of the prism.

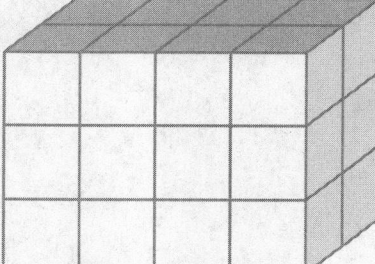

10. A company that manufactures baskets makes them in the shape of a rectangular prism.

Part A: The volume of one basket is 29,750 cubic inches. The basket is 35 inches wide and 34 inches long. Find the height of the basket.

Part B: The company makes two smaller baskets that have a volume of 9,600 cubic inches with a height of 25 inches. Find two possibilities for the length and the width.

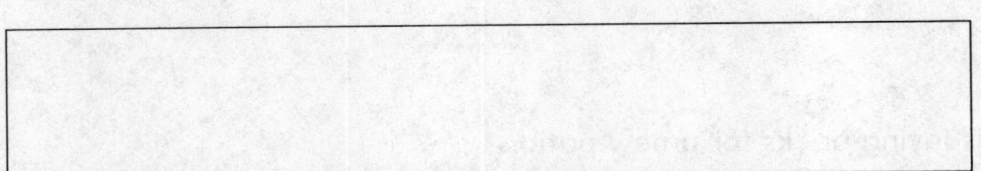

11. Number each of the shapes from 1 to 3 in increasing order according to their volume. Each cube is one cubic centimeter.

      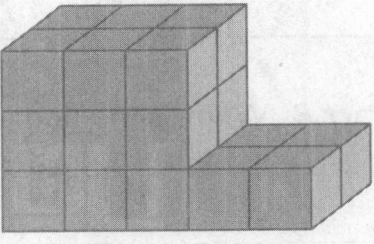

_____   _____   _____

12. Mrs. Huan has 6 books in a stack. The bottom two books are 12 inches long, 1 inch thick, and 6 inches wide. The middle two books are 10 inches long, 1 inch thick, and 5 inches wide. The top two books are 7 inches long, 0.5 inches thick, and 4 inches wide. Find the volume of the stack.

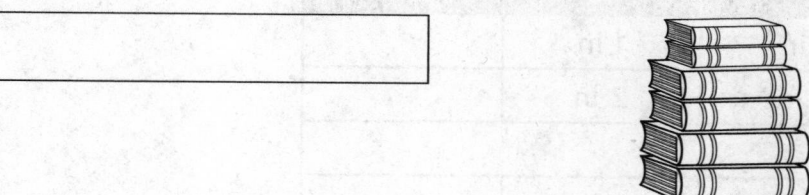

13. A child is stacking blocks in a pyramid design as shown below, but much bigger. There are 66 blocks in all.

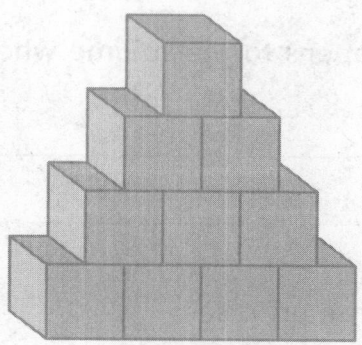

Part A: How many blocks are in the bottom row?

Part B: If each block is a cube with a side length of 2 inches, what is the volume of the construction?

14. Jackson's teacher asked him to look at what happens to the volume of a rectangular prism when every side is doubled.

Part A: Fill in the following chart.

Length	Width	Height	Volume
1 in	1 in	1 in	
2 in	2 in	2 in	
1 in	2 in	3 in	
2 in	4 in	6 in	
3 in	5 in	2 in	
6 in	10 in	4 in	

Part B: Describe what happens to the volume when each side is multiplied by 2.

| |

15. Mark each statement as true or false.

True	False	
☐	☐	All rectangles are squares.
☐	☐	Some parallelograms are rhombuses.
☐	☐	All rhombuses are regular quadrilaterals.
☐	☐	Some trapezoids are parallelograms.

16. An architect is building a house that has 5 faces, 9 edges, and 6 vertices. What is the name for a three-dimensional figure that has these characteristics?

| |

NAME .... DATE ....

SCORE ....

Performance Task

Setting Goals

A factory produces electronic components. The new manager wants to set a goal for how many units will be produced in the upcoming year.

Write your answers on another piece of paper. Show all your work to receive full credit.

Part A

The factory has existed for seven years. The chart below gives the number of components produced by the factory each year.

Year	Components
1	1,432,426
2	1,532,199
3	1,432,501
4	1,570,672
5	1,600,121
6	1,423,411
7	1,531,199

The factory manager needs to put the data in order so that he can make a decision on the next year's goal. Order the data from least to greatest.

Part B

The factory manger asks his assistant manager to give input for the production goal. The assistant manager suggests 1,423,000 units. Explain why this goal may not be appropriate.

Performance Task (continued)

Part C

While the manager is tempted to set a new record by producing more units than have ever been produced in a year, he knows that people are not buying as many components as they used to, and he does not want to make more units than can be sold. He decides to set the goal of producing the third highest number of components in company history. Suggest a goal for the factory manager.

Part D

In researching the company financial reports, the factory manager discovers that the factory must produce at least 1,570,000 units in a year in order to make a profit. Does your goal from **Part C** meet this requirement? If so, explain why. If not, offer the factory manager a new goal that meets *both* requirements.

Part E

The factory manager's supervisor indicates that it is absolutely essential that the total number of units sold in years 6, 7, and the new year 8 be at least 4,500,000. Explain why the goal you gave the factory manager in **Part D** will also meet this new requirement.

NAME .... DATE ....

SCORE ....

Performance Task

Buying Cards

Keena is ordering baseball cards for her store. The boxes come with 24 packs and each pack has 8 cards.

Write your answers on another piece of paper. Show all your work to receive full credit.

Part A

A customer has requested 12 boxes. Keena estimates the order will be about 200 packs. Is her estimate higher or lower than the actual total? Explain your reasoning.

Part B

Suggest a more accurate way of estimating the number of total packs in 8 boxes. Explain why your estimate will be more accurate.

Performance Task *(continued)*

Part C

The store used the following area model to find the total number of packs in 12 boxes. Complete the labeling shown.

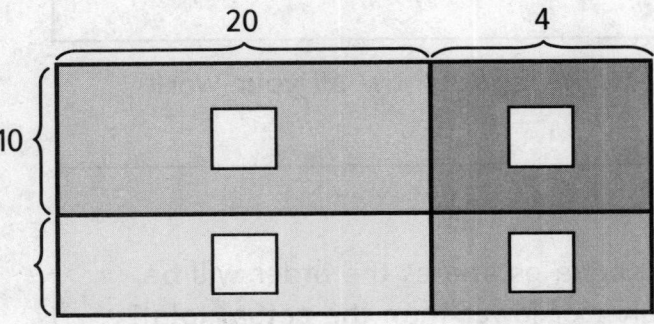

_____ total packs

Part D

How many total cards are in 12 boxes? Explain your reasoning.

Performance Task

Saving for a Bike

Janelle is trying to save money in order to purchase a bike. The bike costs $486. She has three different ways of making money, which are shown in the table below.

Mowing Lawns	$8 per lawn
Walking Dogs	$5 per walk
Washing Cars	$9 per car

Write your answers on another piece of paper. Show all your work to receive full credit.

Part A

Janelle wants to pay for the bike in 6 months. How much does she need to save each month in order to accomplish her goal? Explain.

Performance Task *(continued)*

Part B

In the first two months, Janelle only washes cars. How many cars does she need to wash in order to make her goal for the first two months? Explain.

Part C

In the third and fourth months, Janelle only mows lawns. How many lawns does she need to mow in order to make her goal? Explain the meaning of the remainder.

Part D

How much does Janelle have left to earn? If she only walks dogs for the last two months, how many dogs will she need to walk to make her final goal for buying the bike? Explain

Performance Task

Constructing Frames for an Art Gallery

A woodworker is making picture frames for some rather large paintings for a local art gallery. Each of the paintings has an area of 3,796 square inches. An example of the paintings is shown below:

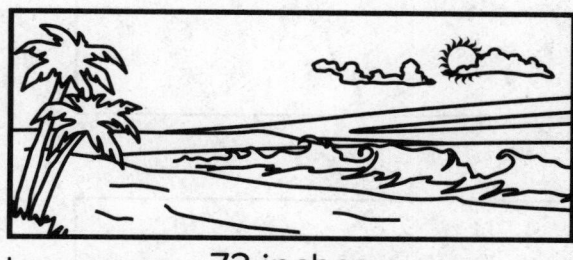

├─── 73 inches ───┤

Write your answers on another piece of paper. Show all your work to receive full credit.

Part A

Find the length of the painting? Explain.

Performance Task *(continued)*

Part B

The woodworker goes to a supply store and finds that boards are only sold in lengths of 5 feet and 8 feet. Are these boards long enough for the project? Explain.

Part C

The woodworker was contracted for 12 frames. He needs to make at least $1,152 in order to make the project worthwhile. How much does he need to charge for each frame?

Part D

Fill in the chart for how many 5-foot boards and 8-foot boards the woodworker will need to fill the order, and fill in the amount of wood left over. Explain.

Board Length	Boards needed for 12 frames	Total inches left over
5 feet		
8 feet		

Performance Task

Planning for a Trip

The Perez family is planning for their summer vacation. The drive will take them three days, and the table shows how many miles they will drive each day.

Day 1	481.23 miles
Day 2	512.94 miles
Day 3	282.22 miles

Write your answers on another piece of paper. Show all your work to receive full credit.

Part A

Estimate the total number of miles that the Perez family will travel for the trip by rounding the number of miles each day to the nearest 10? Show your estimates.

Part B

The family car will get 25 miles per gallon of gas. Use your estimate to determine how many gallons of gas the family will need to buy. Round to the nearest gallon. If gas is $4 per gallon, find out how much money they will need to budget for gas.

Performance Task (continued)

Part C

Based on last year's trip, Mr. Sanchez has planned the following expenses for meals and hotel for one day.

Breakfast	$18.15
Lunch	$22.62
Dinner	$35.67
Hotel	$129.00

The first two days, the family will eat all three meals and will need a hotel room. The third day they will need to eat only breakfast and lunch, and they will not need a hotel room. How much should the family expect to spend on food and lodging? Explain.

Part D

Mr. Sanchez has saved $700 for the trip. Given the estimate for the cost of gas and the expenses for food and lodging, how much can he expect to have left at the end of the three days. Explain.

Performance Task

Making a Healthy Snack Mix

Arundhati wants to make a snack mix with the following healthy ingredients. The total fat, salt, and calories are listed for a serving.

	Almonds	Raisins	Banana Chips
Total Fat	14.25 grams	0.14 grams	10.56 grams
Salt (Sodium)	0.001 grams	3.56 grams	0.002 grams
Calories	168.54	90	155.28

Write your answers on another piece of paper. Show all your work to receive full credit.

Part A

Arundhati will include 2 servings of almonds, 1.5 servings of raisins, and 1.5 servings of banana chips. Find the total fat that is in her mix. Show the expression you used to determine the answer.

Part B

Find the total salt (sodium) that is in Arundhati's snack mix. Show the expression you used to determine the answer.

Performance Task (continued)

Part C

Find the total calories that are in her mix. Show the expression you used to determine the answer.

Part D

If Arundhati has 2 servings of almonds, 1.5 servings of raisins, and 1.5 servings of banana chips, how many total servings does she have in her mix?

Part E

Arundhati wants to calculate the total fat, salt, and calories per serving in her snack mix. Use your answers from **Parts A, B,** and **C,** together with the number of servings from **Part D** to fill in the following table.

Total Fat per Serving	
Salt (Sodium) per Serving	
Calories per Serving	

Performance Task

Planting Trees

Kanona and Latoya have each developed a plan for planting new trees in their local parks. Kanona will plant 2 trees the first year, 4 trees the second year, 8 trees the third year, and so on. Latoya will plant 6 trees the first year, 9 the second year, 12 the third year, and so on.

Write your answers on another piece of paper. Show all your work to receive full credit.

Part A

Describe the pattern for each girl's tree planting plan.

Part B

For each of the first four years, plot the number of trees that each girl plans to plant. Use a different mark for each of the girls.

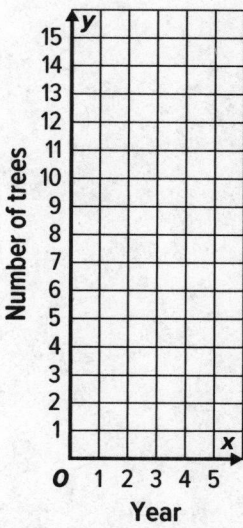

Performance Task (continued)

Part C

For the first couple of years Latoya plants more trees than Kanona. In what year will Kanona plant more trees than Latoya?

Part D

Fill in the table with the total number of trees planted in that year by both girls.

Year 1	
Year 2	
Year 3	
Year 4	

Performance Task

Batting Averages

Winston has been batting for the same baseball team for five years. The table shows how many times he was at bat and how many hits he got.

Year	At Bats	Hits
1	10	3
2	20	7
3	20	4
4	25	8
5	25	10

Write your answers on another piece of paper. Show all your work to receive full credit.

Part A

For each year, what fraction of the times that Winston went to bat did he get a hit? Write your answer as a fraction in lowest terms and as a decimal. The first year is done for you.

Year	Fraction in lowest terms	Fraction as a decimal
1	$\frac{3}{10}$	0.3
2		
3		
4		
5		

Performance Task (continued)

Part B

Put each year in order from Winston's best performance to his worst.

Part C

How many at bats did Winston have over the course of the five years? How many hits did he get in that time? Use this to find the decimal that represents the fraction of the time that Winston got a hit in the five-year period.

Part D

Winston's friend Vince had the exact same batting average in five years. However, Vince only batted 50 times. How many hits did Vince get in this time? Explain.

Performance Task

Triathlon Training

Minh and PJ are training for a triathlon that involves swimming, biking, and running. They spend a week training for each event.

Write your answers on another piece of paper. Show all your work to receive full credit.

Part A

In Week 1 the two athletes are concentrating on swimming. The table shows how many miles each person swam on the given day.

Day	Minh	PJ
Monday	$\frac{1}{3}$ mile	$\frac{1}{4}$ mile
Tuesday	$\frac{1}{2}$ mile	$\frac{1}{2}$ mile
Wednesday	$\frac{3}{4}$ mile	1 mile
Thursday	1 mile	$\frac{2}{3}$ mile
Friday	$\frac{1}{4}$ mile	$\frac{1}{4}$ mile

Which athlete swam further? Explain.

Performance Task (continued)

Part B

In Week 2 the two athletes are concentrating on running and decide to train together. Their goal is to run 60 miles in five days. On Monday they run $10\frac{1}{2}$ miles. On Tuesday they run $13\frac{1}{3}$ miles. On Wednesday they run $11\frac{3}{4}$ miles. On Thursday they run $12\frac{2}{3}$ miles. How far do they have to run on Friday to meet their goal? Explain.

Part C

In Week 3 the two athletes concentrate on biking. Minh's goal is to bike $125\frac{1}{2}$ miles. At the end of the week, he found that he actually biked $125\frac{1}{6}$ miles. PJ's goal was to bike $120\frac{1}{4}$ miles. At the end of the week he found that he actually biked 120 miles. Which athlete was closer to making his goal? Explain.

Performance Task

Creating a Floor Plan

An architect is trying to figure out how to layout a kitchen, a family room, and a living room in a new home. He wants $\frac{1}{2}$ of the floor to be the family room. Once that is marked off, the architect wants $\frac{2}{3}$ of the remaining space to be the kitchen and the rest to be the living room.

Write your answers on another piece of paper. Show all your work to receive full credit.

Part A

A model of the floor is shown below. Divide the floor into areas to find the fraction of the whole that the kitchen will take up. Shade the kitchen. Explain your answer and diagram with a number sentence.

Performance Task (continued)

Part B

The large rectangle is 40 feet by 20 feet. Find the area of the kitchen in square feet. Explain.

Part C

The cost of building the kitchen will be $114 per square foot. Find the cost of constructing the kitchen. Explain.

Performance Task

Comparing Mountains

A scientist wants to compare information on three different mountain hikes. Research ahead of time the height in feet of Mount Everest, Mount McKinley, and Mount Kilimanjaro.

Write your answers on another piece of paper. Show all your work to receive full credit.

Part A

From your research, fill in the heights in feet of the three mountains. Then convert the measurements to miles and feet.

Mountain	Height in Feet	Height in Miles and Feet
Mt. Everest		
Mt. McKinley		
Mt. Kilimanjaro		

Performance Task *(continued)*

Part B

From your research, fill in the heights in meters of the three mountains. Then convert the measurements to kilometers.

Mountain	Height in Meters	Height in kilometers
Mt. Everest		
Mt. McKinley		
Mt. Kilimanjaro		

Part C

The weight of a day pack containing hiking supplies is 20 pounds. How many ounces is this? Show your calculation.

Part D

Each day before climbing, it is recommended that a climber drink 1,500 milliliters of water. How many liters is this? Explain.

Performance Task

Constructing Cereal Boxes

A cereal company is looking to construct a new box for their leading brand of cereal. The box and its dimensions are shown below.

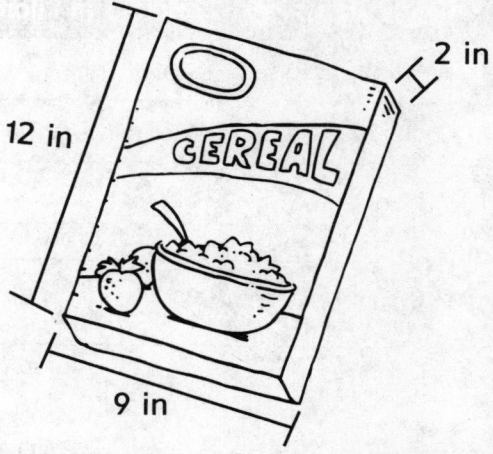

Write your answers on another piece of paper. Show all your work to receive full credit.

Part A

Each box needs to be cut from a flat piece of cardboard. Draw a net for the box and label the length, width, and height.

Performance Task (continued)

Part B

The amount of cardboard needed is measured in square inches. Find the area of the net you drew in **Part A** in order to find the area of cardboard needed to construct a box.

Part C

The box will be filled $\frac{3}{4}$ of the way with cereal. Find the volume of cereal that can be put in each box.

Benchmark Test 1

1. An amusement park admitted 28,512,121 people last year.

   Part A: Fill in the place value chart for the number of people admitted by the park.

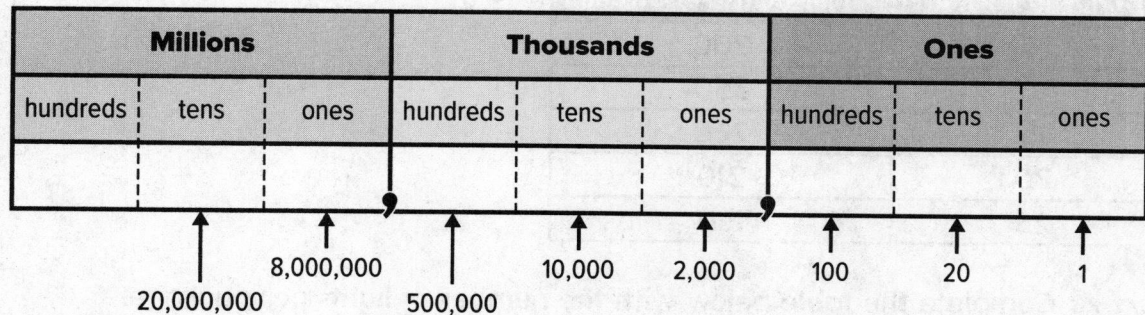

Millions			Thousands			Ones		
hundreds	tens	ones	hundreds	tens	ones	hundreds	tens	ones

   20,000,000   8,000,000   500,000   10,000   2,000   100   20   1

   Part B: Write the number in words.

   Part C: Write the expanded form of the number.

2. Write and solve a division problem that is modeled by the picture.

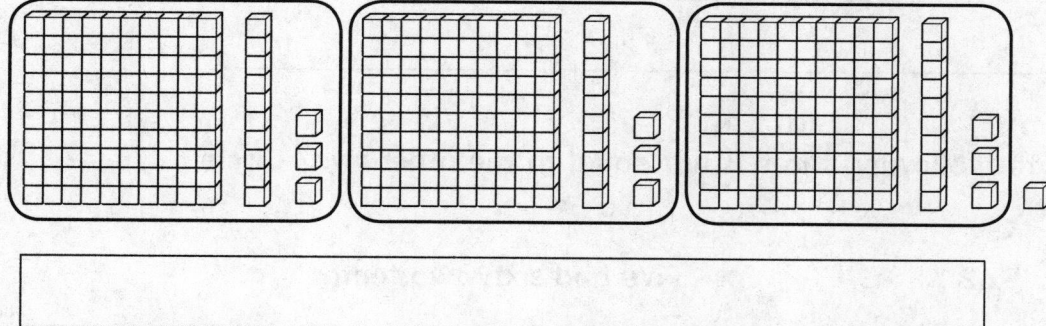

3. The Schmidt family is decorating their house for the winter. Packages of lights cost the same amount, but they have different size strands in them. Mr. Schmidt want to buy the package that has the greatest number of lights.

Light Bulb Packages	
Strands in Package	Light Bulbs Per Strand
5	800
10	350
15	260
20	210
25	160

Part A: Complete the table below with the number of lights per package.

Strands in a Package	5	10	15	20	25
Lights Per Package					

Part B: Which package provides the greatest number of light bulbs?

4. A surveyor is dividing a large plot of land that measures 27,512 square miles. He wants to divide the area into 91 equal regions. Estimate how large each area will be. Show your work.

5. Which of the following three is not equal to the other two? Circle the answer.

5.62   Five and sixty-two tenths

$$5 \times 1 + 6 \times \frac{1}{10} + 2 \times \frac{1}{100}$$

6. The chart shows the cost of several school supplies. Which combinations can you buy with $23?

Pencil	$1
Notepad	$5
Binder	$7
Pen	$2

Yes	No	
☐	☐	2 binders, 1 notepad, and 2 pens
☐	☐	3 binders and 2 pencils
☐	☐	3 notepads, 1 binder, 1 pen, and 1 pencil
☐	☐	1 notepad, 1 binder, 3 pens, and 5 pencils

7. Nine families went on a campout together. The total bill for the weekend supplies was $603. The families will split the bill evenly.

Part A: How much should each family contribute?

Part B: Estimate to check your answer.

8. A local car dealership sells 9,792 cars per year. How many cars does the dealership sell per month?

9. The following chart lists the weight of six packages that came into the post office. Place the weights in order from least to greatest.

3.15 lbs	3.51 lbs	3.05 lbs
5.03 lbs	5.13 lbs	3.015 lbs

[blank answer box]

10. The product of 76 and another number is 15,580. Complete the table to help you estimate the other number.

76 × 100	
76 × 150	
76 × 200	
76 × 250	
76 × 300	

[blank answer box]

11. *Part A:* Fill in the following table with quotients and remainders.

Division Problem	Quotient	Remainder
5,338 ÷ 13		
5,337 ÷ 13		
5,336 ÷ 13		

Part B: What pattern do you notice?

[blank answer box]

12. Jayna drives 17 miles each day to work. She wants to know how many miles she drives in a month. Is there too much information or not enough information to solve this problem? Shade the box next to the correct description. If there is too much information, name the extra information and solve the problem. If there is not enough information, describe what Jayna would need to know to solve the problem.

☐ Too much information   ☐ Not enough information

13. Jenna is trying to use the digits 1, 2, 2, 0, 2, 1 to make a number that is between 210,000 and 220,000. Shade the box next to any answer that is correct.

☐ 122,021   ☐ 210,221

☐ 210,121   ☐ 222,110

14. The Chen family is saving for a vacation to Europe. They need $7,000 for the trip. The family plans to save $312 per month.

Part A: Fill in the partial product diagram to show how much they will save in two years.

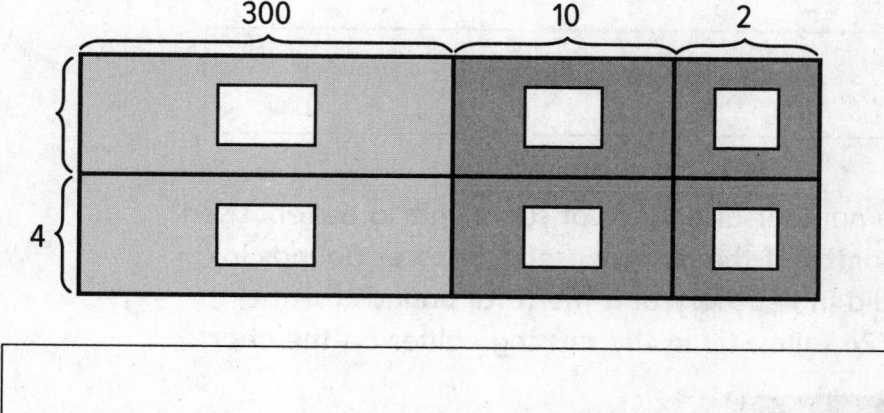

Part B: How much extra money will they have saved?

15. An apple farmer sells apples in bag that hold 7 apples. Her picking crew has picked all the apples that are ready for the weekend sale and begin packaging them into the bags of 7. At the end of the bagging, they have some leftover apples. Shade the boxes next to any number that is a possible remainder, then explain your reasoning.

☐ 0   ☐ 3   ☐ 6

☐ 1   ☐ 4   ☐ 7

☐ 2   ☐ 5   ☐ 8

16. A local charity is storing up large containers of drinking water for emergency purposes. Each container of water costs $19. The charity has $5,147 in donations.

Part A: How many containers of water can the charity buy?

Part B: What is the remainder, and what does it mean?

Part C: Estimate to check your answer. Show your work.

17. The table shows the number of pounds of sugar that a bakery used in three different months. If the bakery used 28 fewer pounds in January than they did in February and the total pounds for the three months was 726 miles, fill in the missing values on the chart.

Month	Pounds of Sugar
January	
February	251
March	

18. The following table shows the number of toothpicks in several different boxes together with the number of boxes in a package. Fill in the table with the missing values.

Number of Toothpicks in a Box	Number of Boxes in a Package	Total Number of Toothpicks in a Package
10^2	85	
10^2		12,500
	176	176,000
10^4	298	

19. A pizza company is open 50 weeks per year. In one year, they sold 8,350 pizzas.

Part A: Fill in the division fact with compatible numbers to estimate the average number of pizzas the company sold per week.

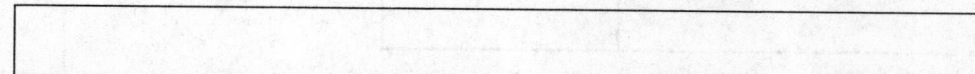 pizzas

Part B: Find the exact average of pizzas sold per week.

Part C: Is your estimate greater than or less than the actual number? Explain how you could have known this ahead of time.

20. Circle any that would not be good ways of estimating $4,924 \div 71$.

$4,900 \div 70 = 70$ $5,000 \div 100 = 50$

$4,000 \div 100 = 40$ $4,000 \div 50 = 80$

Performance Task

A Cross-country Trip

The Perez family is driving from New York City to Los Angeles, but they need to drive through Chicago first to see family. They are trying to plan how many days it will take them to make the trip. Research the distance from New York, NY to Chicago, IL and the distance from Chicago to Los Angeles, CA.

Write your answers on another piece of paper. Show all your work to receive full credit.

Part A

Fill in the table from your research, and round the distances to the nearest hundred miles. Find the total distance the family will travel.

Distance from New York to Chicago	
Distance from Chicago to LA	
Total	

Part B

The Perez car will travel 65 miles per hour. If they plan to drive 8 hours per day, how many miles can they travel in one day? Show your work.

Part C

Find the number of days it will take to drive to Chicago. Find the number of days it will take to drive to Los Angeles. If the Perez family plans to stay in Chicago for 3 days, find the total length of the trip. Show your work.

Benchmark Test 2

1. Round the following number to the nearest hundredth. Write the rounded number in both expanded form and standard form.

$$4 \times 10 + 3 \times 1 + 7 \times \frac{1}{10} + 8 \times \frac{1}{100} + 6 \times \frac{1}{1000}$$

Expanded Form

Standard Form

2. A cook uses 0.4 pounds of butter every morning on croissants. Regroup and shade the models to figure out how much butter he uses in five days.

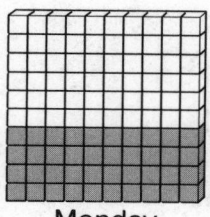
Monday

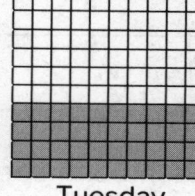
Tuesday

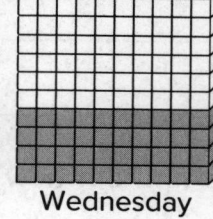
Wednesday

Thursday

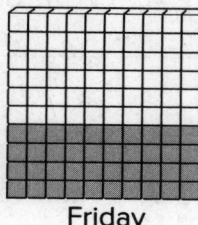
Friday

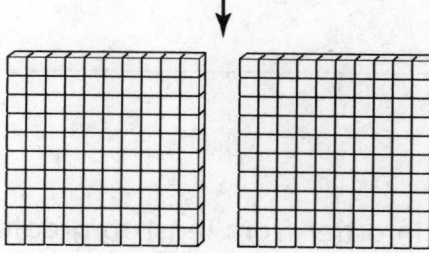

———— pounds

3. A rock is thrown up in the air. The height of the rock in feet after three seconds is $4 \times 3^2 + 6 \times 3 + 12$. Find the height of the rock.

4. Jamal ran 21 miles in 5 days. Circle any of the following that describe the average number of miles Jamal ran per day.

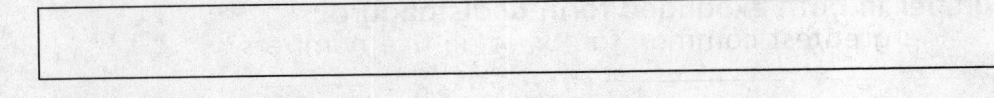

$$\frac{5}{21} \qquad \frac{21}{5} \qquad 4\frac{1}{5} \qquad 5\frac{1}{4}$$

5. A discount book club has a monthly fee of $6. Once you pay the fee, you can buy books for $5 each. In January, Mr. Huan joined the book club and bought 12 books.

Part A: Write an expression for how much Mr. Huan spent.

Part B: Evaluate the expression to find out how much John spent.

6. Put the following numbers in order from least to greatest.

54.03×10^3 $\qquad$ 5.403×10^2 $\qquad$ 0.0543×10^4

7. Consider the following five numbers.

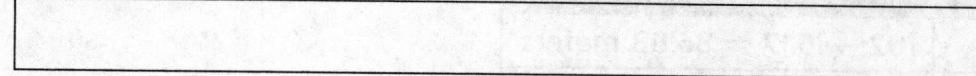

10   15   20   25   30

Part A: What is the greatest common factor of all five numbers?

```
┌─────────────────────────────────────────────────────┐
│                                                       │
└─────────────────────────────────────────────────────┘
```

Part B: Cross out two of the five numbers so that the remaining numbers have a greatest common factor of 10.

8. Write the correct property of addition for each step.

$4.4 + (3.2 + 2.6) + 0$

$= 4.4 + (2.6 + 3.2) + 0$ _____

$= (4.4 + 2.6) + 3.2 + 0$ _____

$= 7.0 + 3.2 + 0$ Addition

$= 10.2 + 0$ Addition

$= 10.2$ _____

9. A factory needs to sell $\frac{2}{3}$ of its inventory in order to make a profit. The fractions below represent the part of the factory's inventory that was sold. Circle any fractions that represent the factory making a profit.

$\frac{7}{11}$ $\frac{9}{13}$ $\frac{10}{16}$

10. An electrician has purchased 102 meters of wire. Each time he wires an outlet in a particular room he uses 15.17 meters of wire. Fill in the following table to figure out how many outlets he can wire and how much wire will be left over.

Outlets Wired	Wire left over
1	102 − 15.17 = 86.83 meters
2	
3	
4	
5	
6	
7	

_____ outlets

_____ meters left over

11. Giovanni was given the following diagram that is supposed to represent a decimal division problem. Write the problem, and find the answer.

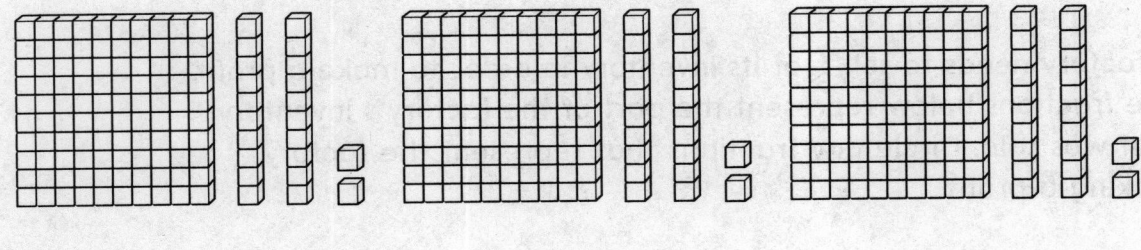

_____ ÷ _____ = _____

12. Circle the pattern that does not belong.

3, 6, 9, 12   2, 4, 8, 16

2, 5, 8, 11   2, 6, 10, 14

13. Match each fraction with its decimal equivalent.

$\frac{11}{25}$ 0.60

$\frac{3}{5}$ 0.75

$\frac{3}{20}$ 0.44

$\frac{3}{4}$ 0.15

14. A bicyclist is riding around town for the morning. The map below shows the places he visits.

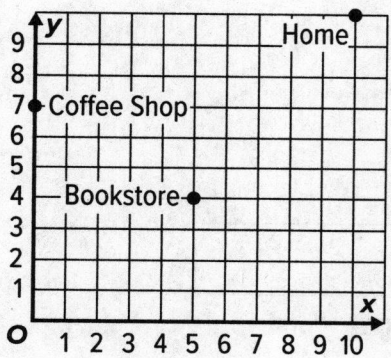

Part A: If the bicyclist starts at home, describe the path he can take to the bookstore, and then to the coffee shop.

Part B: The bicyclist goes directly home from the coffee shop. How many units was his total trip?

15. Shade the box under "Yes" or "No" to indicate whether each problem will require regrouping.

Yes	No	
☐	☐	7.21 + 0.73
☐	☐	29.13 + 20.35
☐	☐	35.05 + 27.05
☐	☐	121.92 + 2.18

16. Glow sticks are sold in packs of various sizes. Which of the three brands is the best buy?

Brand	Number in Pack	Price
A	6	$3.57
B	15	$8.25
C	20	$11.20

17. The table below shows the number of marbles of three different colors that a marble shop has. They want to package the marbles into bags that will have only one color of marble, and they wants to make sure that each bag has the same number of marbles. If all of the marbles are put into bags, what is the greatest number of marbles that could be in each bag?

Pink	75
Red	30
White	60

18. A local fundraising effort managed to raise the following dollar figures in three different activities.

Car Washes	$598
Bake Sales	$243
Yardwork	$102

Part A: Write down the best order in which to add the numbers so that it is easiest to find the total using mental math.

Part B: Find the total amount raised.

19. A soccer coach has marked off an area in the shape of a triangle for some soccer drills. He wants to put a cone at every point that lies *inside* the triangle. Circle any coordinate pair on which the coach should put a cone.

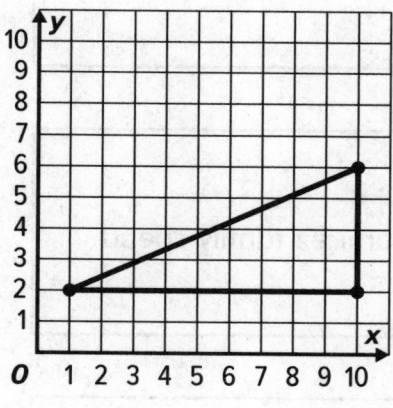

(5, 3)   (8, 4)   (4, 8)   (9, 4)   (3, 7)

20. Circle the expression that is not equal to the others.

988 ÷ 1,000   9.88 ÷ 10

98.8 ÷ 100   98.8 ÷ 10,000

Performance Task

Apple Picking

The Ramirez family is picking apples at an orchard that sells them by the peck. The family came home with 5.25 pecks of apples. Research how many gallons are in a peck.

Write your answers on another piece of paper. Show all your work to receive full credit.

Part A

How many gallons are in a peck? Use this information to figure out how many gallons of apples the Ramirez family bought.

Part B

The family gave 1.5 pecks away to a relative, and gave some more to their neighbors. There are 2.5 pecks left. How many pecks did they give to their neighbor? Explain.

Part C

Each peck of apples cost $5.40. How much did the Ramirez family spend on apples? Explain.

Part D

With the 2.5 pecks of apples that the Ramirez family has left, Mrs. Ramirez intends to make pies. One pie takes 0.25 of a peck. How many pies can she make? Explain.

Benchmark Test 3

1. Juanita has three more than 2 times the number of books than her friend Uma has.

   Part A: If Uma has 6 books, write an expression for the number of books Juanita has. Then find the number of books.

   ┌───┐
   │   │
   │   │
   └───┘

   Part B: If Uma has 7 books, how would the number of books in Juanita's collection compare to your answer in **Part A**? Explain.

   ┌───┐
   │   │
   │   │
   │   │
   └───┘

2. Mr. Ortiz is distributing tennis supplies to his team of 20 players. The extras he will store for future years. Write an expression for the number of supplies each player receives and then evaluate each expression.

Supplies	Expression	Each Player Receives
95 tennis balls		
61 racket grips		
1,200 bottles of water		
24 t-shirts		

3. Which of the following can be modeled by the division expression 550 ÷ 5? Choose all that apply.

   A. 550 dollars distributed evenly to 5 groups

   B. 5 points distributed evenly 550 times

   C. 550 pounds distributed evenly into bags of 5 pounds each

   D. 550 feet per step for 5 steps

4. Compare $\frac{7}{10}$ and $\frac{7}{100}$.

Part A: Shade the decimal models to show each fraction. Then write each as a decimal.

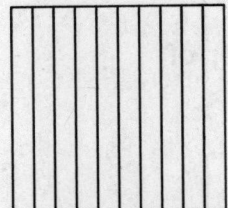 _____    _____

Part B: Compare the two decimals. Use >, <, or =. Explain.

5. A farmer measured a sample of his corn plants after two weeks and recorded the following measurements in inches: $4\frac{1}{2}$, 5, $6\frac{3}{4}$, $7\frac{1}{4}$, $4\frac{1}{2}$, $6\frac{1}{4}$, 7, $6\frac{3}{4}$, 7, $6\frac{3}{4}$.

Part A: Use the line plot to record the measurements.

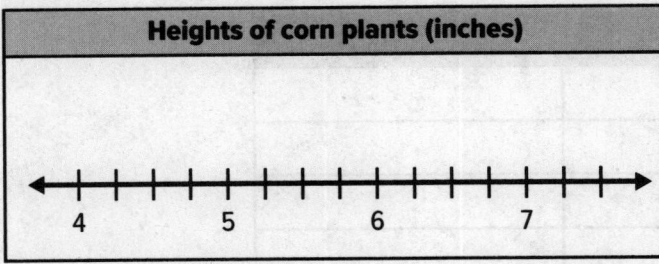

Part B: What is the sum of the heights of all the corn plants?

6. Each dog in a kennel needs 8 pounds of food for their upcoming stay. If the kennel has 225 pounds of food, will there be enough to accommodate 20 dogs? If so, how many more dogs can be accommodated? Explain your reasoning.

7. Draw the decimal points on each number on the left side of the equation so that the difference is correct as shown.

$$3\ 2\ 5 - 4\ 2 = 28.3$$

8. Shanna is delivering papers on her morning route. She starts at (0, 0), and three of her houses are at points 3 right, 6 up; 2 up, 1 right; and 5 right, 3 up. Use the grid to draw and label the points where the houses are located.

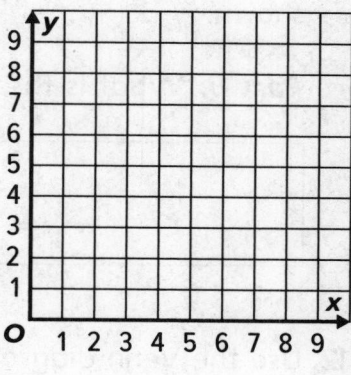

9. While dividing numbers with zeros at the end, Ethan notices a certain pattern. His results are shown in the table.

Expression	Quotient
100 ÷ 5	20
1000 ÷ 50	20
10 ÷ 5	2
100 ÷ 50	2

Part A: What pattern does he recognize?

Part B: Using this pattern, what is the result of 10,000 ÷ 50?

10. In a competition for the tallest stack of balanced blocks, the top four heights were recorded as follows: 2.9 m, 2.77 m, 2.81 m, 2.84 m. Place a dot on the number line for each given height and label.

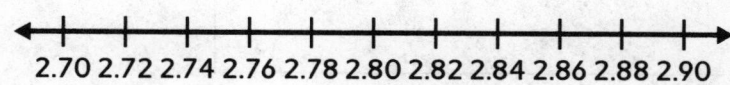

2.70 2.72 2.74 2.76 2.78 2.80 2.82 2.84 2.86 2.88 2.90

11. A carpenter is cutting a piece of wood that is 1 yard by 1 yard. He wants one side to be $\frac{3}{8}$ of a yard and one side to be $\frac{1}{4}$ of a yard.

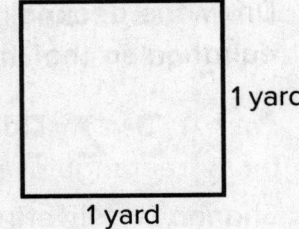

1 yard

1 yard

Part A: Model the desired cut on the square of wood shown.

Part B: What is the area of the wood?

12. Use the Venn diagram to sort the shapes. Draw a line from the shape to the correct area of the diagram.

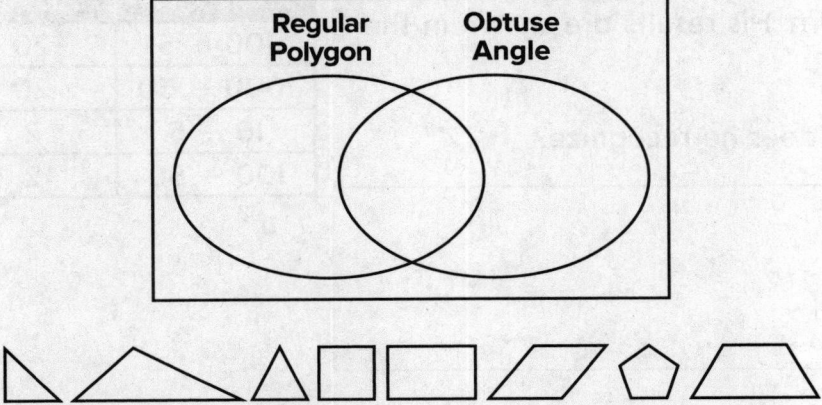

13. A group of friends earned $5,520 doing yard work around town for a year. They decided to give one third of the money to a charity. Then each friend received an equal portion of the money, but less than what was given to charity. Were there 2, 3, or 4 friends? How much money did each friend receive? Explain your reasoning.

14. Julian has 12 unit blocks. He needs to create rectangular prisms.

Part A: Use the table shown to create as many rectangular prisms as you can.

Part B: As long as the prism has the same three numbers for the sides, it is considered to be the same. How many unique prisms can you create?

Length	Width	Height

15. Draw lines between equivalent numbers.

$1\frac{6}{7}$ $\frac{3}{4}$

$3\frac{2}{5}$ $\frac{2}{3}$

$\frac{16}{24}$ $\frac{17}{5}$

$\frac{18}{24}$ $\frac{13}{7}$

16. The order for a collection of artwork is shown. What is the total area of canvas needed to for the order of all of the artwork?

Artwork Order		
Quantity	Dimensions (in.)	Total Square Inches
6	$12\frac{3}{4} \times 8$	
2	$10\frac{3}{8} \times 10$	
8	$15\frac{1}{2} \times 9$	

Total = _____

17. Which of the following use the proper order of operations? Select all that apply.

- ☐ $20 - (6 + 5 \times 2) = 38$
- ☐ $4 \times [8 - (10 \div 2)] = 12$
- ☐ $15 - 1 + 3 = 11$
- ☐ $4^2 + 2 \times 5 = 26$
- ☐ $2 \times (24 + 1) = 50$

18. There are 10 milligrams in every centigram, 100 milligrams in every decigram, and 1000 milligrams in every gram.

Part A: How many centigrams are there in a gram? Explain.

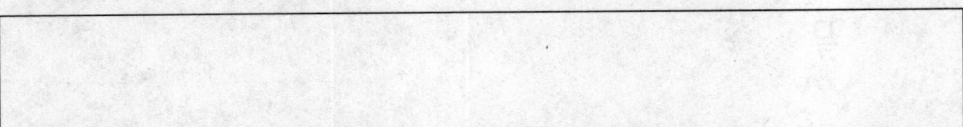

Part B: How many decigrams are there in a gram? Explain.

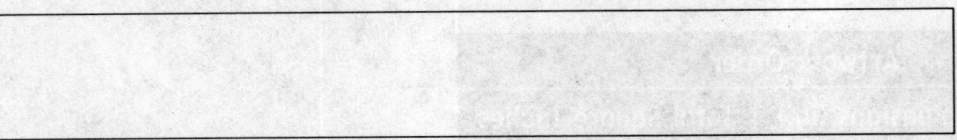

Part C: How does a gram compare to a decigram? Explain.

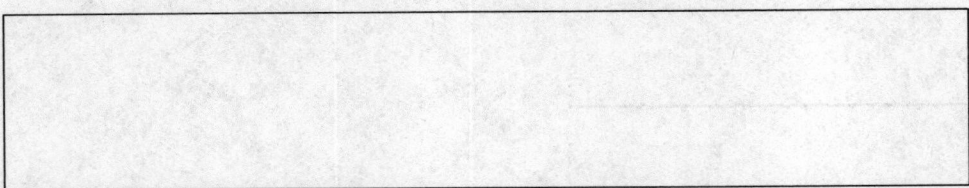

19. While trying to add lengths for marking off a garden plot, Mrs. Shen calculated that $\frac{3}{5} + \frac{1}{2} = \frac{4}{7}$. Use estimation to determine if Mrs. Shen is correct.

20. Martina compared numbers with similar digits. Using mathematical language, explain how each set of numbers is different.

A	12,678 and 2,678
B	576 and 57.6
C	49 and 049

21. At the end of a night, a cashier empties the registers of the $1, $10, and $100 bills. There are 40 bills total. Which of the following would make the least amount of money? Explain how you solved the problem.

Possible Bill Combinations		
$1 Bills	$10 Bills	$100 Bills
6	34	0
4	35	1
16	22	2
3	36	1

22. Pianos have 88 keys. If a company produces 620 keys, how many pianos can be produced? What does the remainder mean in this case?

23. Belinda's work is shown for a recent test on fraction operations.

$$7 \div \frac{1}{5} = \frac{7 \div 1}{5} = \frac{7}{5}$$

Part A: What mistake did Belinda make while dividing?

Part B: What is the correct quotient?

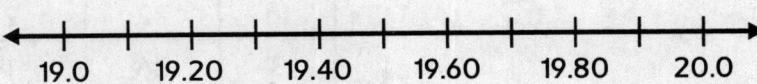

24. A long jumper jumped 19.569 feet.

Part A: Round this number to the nearest hundredth.

Part B: Place rounded number on the number line shown.

19.0   19.20   19.40   19.60   19.80   20.0

25. Explain how you model 0.6 × 0.5 using a decimal grid. How would this vary from 1.6 × 0.5?

26. While trying to solve 4.21 + 0.52, Raj found the sum to be 42.62. Explain why this is or is not a valid answer using estimation.

27. A baker gradually added flour to a mixing bowl for bread. He started with 3 cups of flour. He then added 1.5 cups. The third time he added 5 ounces less than the second time. Finally he added 5 ounces. How many cups of flour did he add in all? Explain your reasoning.

28. A piece of wire is 2.4 meters in length. An electrician can create 4 equal sized wires from this piece or 6 equal sized wires from this piece. Use the number lines below to model each option.

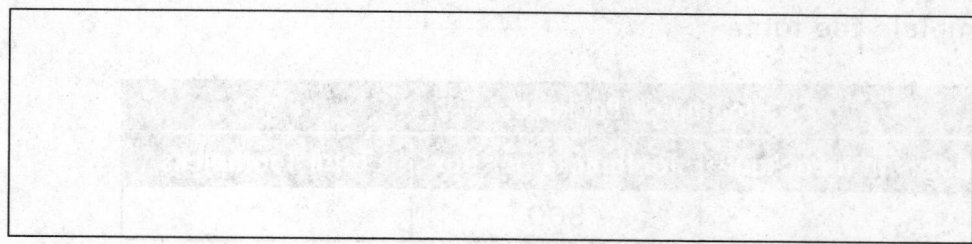

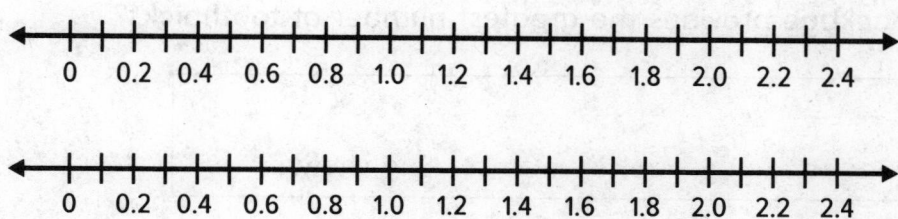

29. Juan went on four runs over the past week. The first two measured $\frac{3}{4}$ mile and $1\frac{1}{4}$ miles. The third and fourth measured $\frac{7}{8}$ mile and $1\frac{3}{4}$ miles.

Part A: Find the sum of the first two, then the sum of the last two.

Part B: Describe how the adding processes differed.

Part C: What is the total of all four runs?

30. A restaurant owner is buying packages of toothpicks. Each package comes with a certain number of boxes that contain the toothpicks. Because they all cost roughly the same price, he decides he wants the package that provides the most toothpicks.

Part A: Complete the table

Toothpick Packages		
Number of Boxes in a Package	Toothpicks per Box	Total Toothpicks
2	800	
3	450	
5	350	
10	160	

Part B: Which package provides the greatest number of toothpicks?

31. You have a recipe for a fruit smoothie, but you want to increase the recipe so that it feeds $3\frac{1}{2}$ times the original number of people. Complete the table by writing the new measurements.

Original Recipe	$3\frac{1}{2}$ Times Recipe
1 cup bananas	
$2\frac{1}{2}$ cups strawberries	
$\frac{3}{4}$ cup of milk	
2 tablespoons of honey	

Performance Task

Building a Corn Bin

An engineer is preparing plans for building a structure to hold spare corn for a farmer. The structure will be a rectangular prism that is $12\frac{1}{2}$ feet wide, 16 feet long, and $6\frac{1}{2}$ feet tall.

Write your answers on another piece of paper. Show all your work to receive full credit.

Part A

The main structure will be built with solid steel beams. Determine how many beams of each length the engineer will need for the frame, and plot the lengths on a line plot.

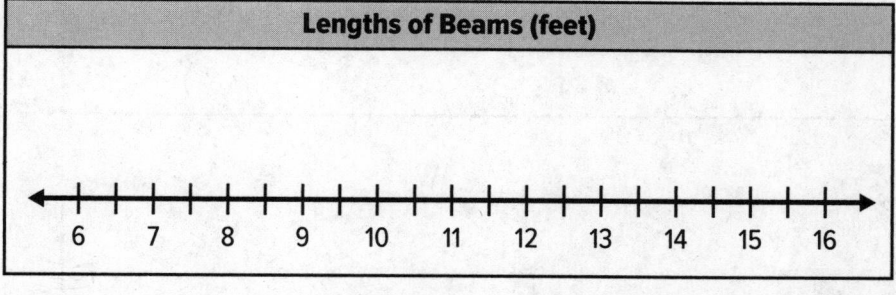

Part B

Beams only come in lengths of 8 feet, 15 feet, and 20 feet. The engineer will cut the lengths he needs for the bin from these sizes. Each beam can only be cut one time. Find the length of beam in inches that will be left over after the engineer makes his cuts. Show your calculation.

Performance Task (continued)

Part C

Find the volume of the bin in cubic feet. Show your calculation.

Part D

Suppose the farmer changes the project requirements and wants the bin to be 25 feet wide instead of $12\frac{1}{2}$ feet wide. How will the volume of the new bin compare to the volume of the original bin? Explain your reasoning.

Benchmark Test 4

1. Mr. Li is buying packages of notebooks. Because they all cost roughly the same price, he decides he wants the package that provides the most sheets of paper.

Notebook Package Sizes	
Number of Notebooks	Sheets per package
1	800
3	350
6	250
10	150

Part A: Complete the table below with the number of total sheets per package.

Notebooks	Total Sheets
1	
3	
6	
10	

Part B: Which package provides the most total sheets of paper? Justify your response.

2. Compare each number to 85.2. Use the symbols <, >, or =.

85.20 ◯ 85.2

852 + 0.2 ◯ 85.2

eighty-five ones and two tenths ◯ 85.2

852 hundredths ◯ 85.2

3. In a competition for the tallest sunflower, the top four heights
   were recorded as follows: 3.9 m, 3.88 m, 3.82 m, 3.76 m.

   Place a dot on the number line and label for each given distance.

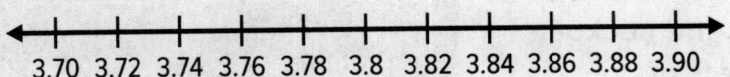

   3.70 3.72 3.74 3.76 3.78 3.8 3.82 3.84 3.86 3.88 3.90

4. Julian compared numbers with similar digits. Using mathematical
   language, explain how each set of numbers is different.

A	13.542 and 35.42
B	781 and 78.1
C	1.2 and 01.20

5. There are 10 millimeters in every centimeter, 100 millimeters in every decimeter, and 1000 millimeters in every meter.

   Part A: How many centimeters are there in a meter? Explain.

   Part B: How many decimeters are there in a meter? Explain.

   Part C: How does a meter compare to a decimeter? Explain.

6. Compare $\frac{5}{10}$ and $\frac{5}{100}$.

   Part A: Shade the decimal models to show each fraction. Then write each as a decimal.

   _____   _____

   Part B: Compare the two decimal fractions. Use >, <, or =. Explain.

7. While trying to add together the widths of some boards for a deck he is building, Haj said that $\frac{4}{7} + \frac{1}{2} = \frac{5}{9}$. Use estimation to determine if Haj is correct.

8. An academic team won a scholarship of $5,856. One fourth of the money was first given to a charity. Then each team member received an equal portion of the scholarship money, but less than what was given to charity. Were there 2, 3, or 4 team members? How much money did each team member receive? Explain your reasoning.

9. Which of the following can be modeled by the division expression 650 ÷ 5? Choose all that apply.

   A. $650 distributed evenly to 5 groups

   B. 5 points distributed evenly 650 times

   C. 650 ounces distributed evenly into beakers of 5 ounces each

   D. 650 millimeters per step for 5 steps

10. Each student needs 11 pencils for the school year. If the school started with a box of 1,325, would there be enough for a school of 120? If so, how many more students can be supplied pencils? Explain your reasoning.

11. Ms. Marcella is distributing school supplies to her classroom of 30 students. Write an expression for the number of supplies each student receives and then evaluate each expression.

Supplies	Expression	Each Student receives
100 folders		
65 highlighters		
1,000 sheets of paper		
34 calculators		

12. While dividing two numbers with zeros at the end, Eugenio notices a certain pattern. His results are shown in the table.

Expression	Quotient
100 ÷ 2	
1000 ÷ 20	
10 ÷ 2	
100 ÷ 20	

Part A: What pattern does he recognize?

Part B: Using this pattern, what is the result of 10,000 ÷ 20?

13. If 365 work days are to be split among 12 employees evenly, how can you rewrite this as a fraction? What does the fraction represent?

14. A grasshopper jumped 6.434 centimeters.

Part A: Round this number to the nearest hundredth.

Part B: Place the rounded number on the number line shown.

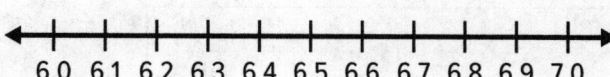

6.0 6.1 6.2 6.3 6.4 6.5 6.6 6.7 6.8 6.9 7.0

15. While trying to add 1.43 + 0.71, Kevin found the sum to be 15.01. Use estimation to explain why this is or is not a valid answer.

16. Draw the decimal points on each number on the left side of the equation so that the difference is correct as shown.

$$2 \ 1 \ 4 \ - \ 2 \ 3 \ = \ 19.1$$

17. A strip of cloth is 1.4 meters in length. A tailor can create 7 equal sized strips from this piece or 2 equal sized strips from this piece. Use the number lines below to model each option.

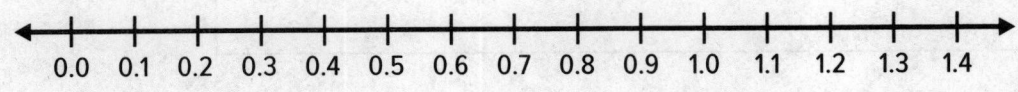

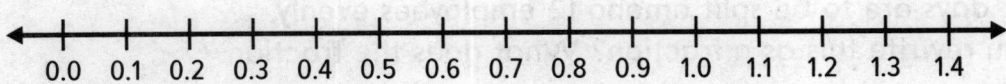

18. Aaron's work is shown for a recent operations on fractions test.

$$4 \div \frac{1}{4} = \frac{4 \div 1}{4} = \frac{4}{4} = 1$$

Part A: What mistake did Aaron make while dividing?

Part B: What is the correct quotient?

19. Explain how you model 0.7 × 0.8 using a decimal grid. How would this vary from 1.7 × 0.8?

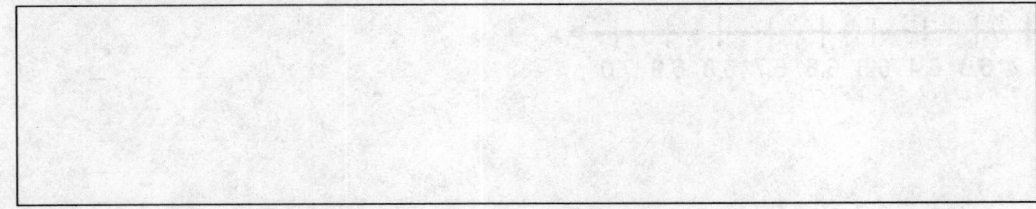

20. Which of the following uses the proper order of operations? Select all that apply.

☐ $3 \times (54 + 7) = 183$

☐ $12 - 5 + 2 = 5$

☐ $3 \times [5 - (6 \div 2)] = 6$

☐ $3^2 + 7 \times 4 = 37$

☐ $24 - (7 + 3 \times 2) = 40$

21. John is four more than 3 times his daughter's age.

Part A: If John's daughter is 10 years old, write an expression for John's age. Then calculate his age.

Part B: If John's daughter is 9, how would John's age compare to your answer in *Part A*? Explain.

22. Draw lines between equivalent fractions.

$1\frac{3}{5}$ $\frac{13}{5}$

$2\frac{3}{5}$ $\frac{8}{5}$

$\frac{8}{32}$ $\frac{1}{4}$

$\frac{6}{32}$ $\frac{3}{16}$

23. A fire fighting robot needs to be programmed to go where the fire is. (0, 0) is considered the entrance to the room, and the fires are at points 2 right, 8 up; 2 up, 2 right; and 6 right, 2 up. Use the grid to draw and label the points where the fires are located.

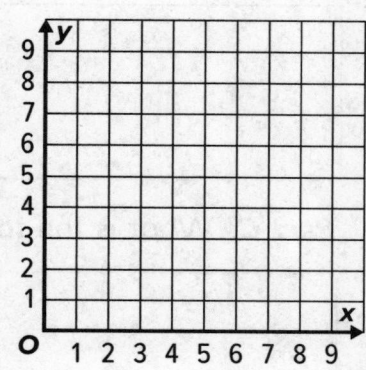

24. A tailor has a piece of cloth that is I yard wide by I yard long. From this he will cut a piece of cloth that is $\frac{5}{8}$ yard wide by $\frac{3}{4}$ yard long.

1 yard

1 yard

Part A: Model the desired cut on the square of fabric shown.

Part B: What is the area of the fabric? Write an equation.

25. Use the Venn diagram to sort the shapes. Draw the line from the shape to the correct area of the diagram that describes it.

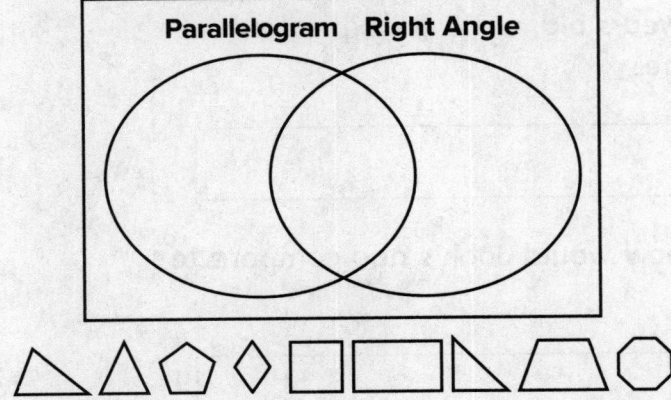

Parallelogram   Right Angle

26. Marcus is adding $\frac{2}{3}$ cup and $1\frac{1}{3}$ cups of cereal together in a bowl. In another bowl, he adds $\frac{5}{8}$ cup and $1\frac{1}{4}$ cups of dried fruit.

Part A: Find the sum of the contents of each bowl.

Part B: Describe how the adding processes differed between finding the two sums.

Part C: What is the total of contents in both bowls together?

27. The order for the Grade 5 class photo is shown. What is the total square footage of photo paper needed to print all the photos? Write an equation to show your work.

Grade 5 Class Photos		
Quantity	Dimensions (in.)	Total Square Feet
7	$7\frac{3}{4} \times 5$	
3	$3\frac{3}{8} \times 4$	
8	$8 \times 10\frac{1}{2}$	

Total =

28. While measuring their bean plants after two weeks, the class recorded the following measurements in inches: $5\frac{1}{2}$, 6, $5\frac{3}{4}$, $6\frac{1}{4}$, $5\frac{1}{2}$, $6\frac{1}{4}$, 6, $5\frac{3}{4}$, 6, $5\frac{3}{4}$.

Part A: Use the line plot to record the measurements.

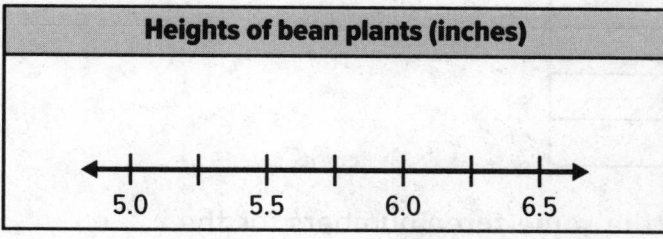

Heights of bean plants (inches)

5.0   5.5   6.0   6.5

Part B: What is the total height of all the bean plants together?

29. You have a recipe for pancakes, but you want to increase the recipe so that it feeds $2\frac{1}{2}$ times the original number of people. How much of each ingredient do you need now?

Original Recipe	$2\frac{1}{2}$ Times Recipe
I cup flour	
$\frac{2}{3}$ cup sugar	
$\frac{3}{4}$ tablespoons baking powder	
$\frac{1}{2}$ teaspoons salt	

30. A cat jumped 3 times. Starting at the porch, he jumped 3 feet. Then he jumped 5 inches less than the first jump. Finally, he jumped $1\frac{1}{2}$ feet further than his second jump. What were the lengths of his jumps? How far from the porch is the cat now?.

31. Kellen has 20 unit blocks. He needs to create rectangular prisms.

Part A: Use the table shown to create as many rectangular prisms as you can.

Length	Width	Height

Part B: As long as the prism has the same three numbers for the sides, it is considered to be the same. How many unique prisms can you create?

Performance Task

Building a Sandbox

You are building a sandbox for your little brother. The width of all the spare boards you have is 1 foot. The lengths of these boards are shown in the table below.

Board Length (ft)		
$5\frac{1}{2}$	6	$4\frac{1}{2}$
6	$5\frac{1}{2}$	$5\frac{1}{2}$
$5\frac{1}{2}$	5	$5\frac{1}{2}$
5	5	6

Part A

Create a line plot using the measurements shown.

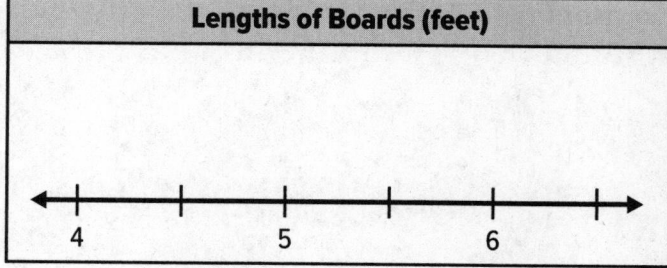

Lengths of Boards (feet)

Performance Task (continued)

Part B

The measure of the sandbox will be I foot high, 6 feet wide, and $5\frac{1}{2}$ feet long. Draw and label the sandbox. What is the total length of the boards that will not be used?

Find the volume of the sand you need to buy, in cubic inches, to fill the finished sandbox to the top.

Part C

Suppose you want the sandbox to be 3 feet wide instead of six feet wide. How will the new volume compare to the volume of the sandbox in **Part B**? Explain your reasoning.

Assessment Item Types

Assessment Types

Copyright © McGraw-Hill Education. Permission is granted to reproduce for classroom use.

Assessment Item Types

In the spring, you will probably take a state test for math that is given on a computer. The problems on the next few pages show you the kinds of questions you might have to answer and what to do to show your answer on the computer.

Selected Response means that you are given answers from which you can choose.

Selected Response Items

Regular multiple choice questions are like tests you may have taken before. Read the question and then choose the <u>one</u> best answer.

Multiple Choice

Four yards of fabric will be cut into pieces so that each piece is thirteen inches long. How many pieces can be cut?

- ☐ 6 pieces with 2 inches left over
- ☐ 7 pieces with 1 inch left over
- ☐ 10 pieces with 2 inches left over
- ☑ 11 pieces with 1 inch left over

ONLINE EXPERIENCE Click on the box to select the one correct answer.

HELPFUL HINT Only one answer is correct. You may be able to rule out some of the answer choices because they are unreasonable.

▶ **Try On Your Own!**

Four boxes to be mailed are weighed at the post office. Box A weighs 8.22 pounds, Box B weighs 8.25 pounds, and Box C weighs 8.225 pounds. Box D weighs less than Box C but more than Box A. How much could Box D weigh?

- ☐ 8.22 pounds
- ☑ 8.224 pounds
- ☐ 8.226 pounds
- ☐ 8.23 pounds

Sometimes a multiple choice question may have more than one answer that is correct. The question may or may not tell you how many to choose.

Multiple Correct Answers

Select all values that are equivalent to 332 ounces.

- ☑ 2 gallons, 76 ounces
- ☑ 20 pints, 12 ounces
- ☐ 22 pints, 8 ounces
- ☐ 41 cups, 5 ounces
- ☑ 41 cups, 4 ounces

▶ **Try On Your Own!**

Select all statements that are true.

- ☑ All rhombuses are parallelograms.
- ☐ All trapezoids are parallelograms.
- ☐ All rectangles are trapezoids.
- ☑ All squares are rectangles.

ONLINE EXPERIENCE Click on the box to select it.

HELPFUL HINT Read each answer choice carefully. There may be more than one right answer.

Assessment Item Types   vii

Assessment Item Types   viii

Copyright © McGraw-Hill Education. Permission is granted to reproduce for classroom use.

Copyright © McGraw-Hill Education. Permission is granted to reproduce for classroom use.

Grade 5 · Assessment Item Types

189

Another type of question asks you to tell whether the sentence given is true or false. It may also ask you whether you agree with the statement, or if it is true. Then you select yes or no to tell whether you agree.

ONLINE EXPERIENCE Click on the box to select it.

HELPFUL HINT There is more than one statement. Any or all of them may be correct.

Multiple True/False or Multiple Yes/No

Determine whether each polygon shown is also a rhombus. Select Yes or No for each polygon.

Yes   No

□   ▨

▨   □

▨   □

□   ▨

Try On Your Own!

Select True or False for each comparison.

True   False

▨   □   200 centimeters > 1.5 meters

□   ▨   36 inches > 2 yards

▨   □   1 gallon > 12 cups

▨   □   2 miles < 3,500 yards

You may have to choose your answer from a group of objects.

ONLINE EXPERIENCE Click on the figure to select it.

HELPFUL HINT On this page you can draw a circle or a box around the figure you want to choose.

Click to Select

A rectangular prism has a length of 12 centimeters, a width of 8 centimeters, and a height of 32 centimeters. Which equations could be used to find the volume of the rectangular prism in cubic centimeters?

$12 + 8 + 32 = V$ $12 \times 8 \times 32 = V$

$(12 + 8) \times 32 = V$ $(32 \times 8) \times 12 = V$

$96 \times 32 = V$ $18 \times 32 = V$

Try On Your Own!

Select all expressions that are equal to $5\frac{1}{3}$.

$16 \times \frac{1}{3}$ $2\frac{1}{3} \times 2\frac{2}{3}$ $32 \times \frac{1}{6}$

$15 \times \frac{1}{3}$ $8 \times \frac{2}{3}$ $3\frac{1}{3} \times 2$

When no choices are given from which you can choose, you must create the correct answer. One way is to type in the correct answer. Another way may be to make the correct answer from parts that are given to you.

Constructed-Response Items

Fill in the Blank

The table shows the number of laps Tammi ran around the track each day. Complete the table if the pattern continues.

Day	Laps
1	4
2	7
3	10
4	13
5	16
6	19
7	22

ONLINE EXPERIENCE You will click on the space and a keyboard will appear for you to use to write the numbers and symbols you need.

HELPFUL HINT Be sure to provide an answer for each space in the table.

Try On Your Own!

Sasha planted a garden in her backyard that is 32 square feet in area. If the length was 8 feet, how many inches wide was the garden?

48

Assessment Item Types   xi

Sometimes you must use your mouse to click on an object and drag it to the correct place to create your answer.

Drag and Drop

Drag one expression to each box to make the statements true.

Subtract 3 from 9 and then add 2.	=	$9 - 3 + 2$
Add 3 and 9 and then subtract 2.	=	$3 + 9 - 2$
The sum of 3 and 2 is subtracted from 9.	=	$9 - (3 + 2)$

$3 + 9 - 2$

$9 - (3 + 2)$

$9 - 3 + 2$

ONLINE EXPERIENCE You will click on an expression and drag it to the spot it belongs.

HELPFUL HINT Either draw a line to show where the expression goes or write the expression in the blank.

Try On Your Own!

Order from least to greatest by dragging each number to a box.

3.045	3.109	3.103	3.17	3.059
3.016	3.045	3.059	3.103	3.016

3.109	3.17

xii   Assessment Item Types

Some questions have two or more parts to answer. Each part might be a different type of question.

Multipart Question

Connor is filling a 15-gallon wading pool with water. On his first trip he carried $3\frac{7}{12}$ gallons of water. He carried $3\frac{1}{3}$ gallons on his second trip, and $2\frac{1}{2}$ gallons on his third trip.

Part A: How much water did Connor carry to the wading pool on trips 1, 2, and 3?

$9\frac{5}{12}$ gallons

Part B: How many more gallons will Connor need to carry to the wading pool until it is filled?

$5\frac{7}{12}$ gallons

Try On Your Own!

This table shows the three different ways that apples are sold at Donaldson's Fruit Farm in the fall.

Package Type	Amount in the Package
Bag	12 apples
Box	8 bags
Crate	15 boxes

Part A: Select the expression that can be used to determine the number of bags of apples that are in a crate of apples.

- ☑ 8 × 15
- ☐ 12 + 8 + 15
- ☐ 12 × 8 × 15
- ☐ 12 × 8

Part B: The label on a bag of apples states that it contains 1.25 pounds of apples. What is the total weight, in pounds, of the bags of apples in one crate?

One crate of apples weighs [150] pounds.

NAME

DATE

SCORE

Countdown: 20 Weeks

1. Ben is playing a game with his friend Keke. The person who is able to compose the greatest six-digit number wins. After six spins, they have the numbers 5, 8, 5, 0, 6, and 3. The table shows the numbers they each composed.

| Ben | 860,553 |
| Keke | 865,530 |

Part A: Who composed the greatest six-digit number? Explain.

Keke. Sample answer: When comparing the value of numbers, each digit to the left is 10 times greater than the digit to its right. So, the larger digits should be in the greatest place-value positions all the way down to the ones position. The zero in Ben's number should be in the ones position.

Part B: Shade the box in front of each number whose value falls between Keke's number and Ben's number.

□ 860,550 □ 863,492 □ 806,553
■ 865,529 ■ 864,942 □ 860,552

Part C: Place the numbers from Part B in order from *least to greatest.*

806,553; 860,550; 860,552; 863,492; 864,942; 865,529

2. Arjun is putting his football cards in order from lowest number to highest number. He only has two cards in the 300s. One card is number 361. The other card is damaged, but he can read part of the number 3__4. Which actual card numbers would be greater than 361? Which actual card numbers would be less than 361?

304, 314, 324, 334, 344, 354 are less than 361. 364, 374, 384, 394 are all greater than 361.

3. Choose two different ways you could write the number represented by the grids shown? Explain your reasoning.

Sample answer: 3.24, $3 + \frac{2}{10} + \frac{4}{100}$. The grids have 3 whole grids, two columns, and 4 extra blocks, which represent ones, tenths, and hundredths respectively.

4. Triple jumps at a track meet have the following distances: 16.08 m, 16.1 m, 16.02 m, 16.20 m.

Part A: Place a dot on the number line and label for each given distance.

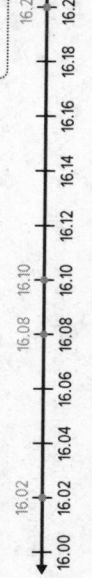

16.00 16.02 16.04 16.06 16.08 16.10 16.12 16.14 16.16 16.18 16.20

Part B: Which jump was the longest?

16.20 m

5. Compare each number to 2.15. Use the symbols <, >, or =.

2.150 (=) 2.15

$2 + \frac{15}{10}$ (>) 2.15

2 ones and 15 thousandths (<) 2.15

215 hundredths (=) 2.15

Countdown

NAME

DATE

SCORE

Countdown: 19 Weeks

1. Julian was taking notes for a report on the U.S. population. When reading his notes later, he found he couldn't read all the numbers. He did remember the following information.

A. The smallest place-value position is 6.

B. The number in the hundred thousands place has a value that is $\frac{1}{10}$ the value of the number in the millions place.

C. The value of the number 7 is 7 thousands.

Using the hints from above, write the missing digits in the chart.

3	1	8	8	5	7	0	5	6

2. Julian compared numbers with similar digits. Using mathematical language, explain how each set of numbers is different.

A	53,671 and 52,671
B	354 and 3.54
C	152 and 0152

Sample answer:

A: The second number is 1,000 less than the first number.

B: The first number is a whole number. It represents 354 ones. The second number is a decimal fraction. It represents 3 ones and 54 hundredths of 1 whole.

C: Both numbers have the same value. When a zero is placed as the first digit in a whole number, it has no value.

---

3. There are 10 years in every decade, 100 years in every century, and 1000 years in every millenium.

Part A: How many decades are there in a millenium? Explain.

Sample answer: 100 decades in a millennium; One decade is $\frac{1}{100}$ of a millennium.

Part B: How many centuries are there in a millenium? Explain.

Sample answer: 10 centuries in a millennium; One century is $\frac{1}{10}$ of a millennium.

4. Compare $\frac{3}{10}$ and $\frac{3}{100}$.

Part A: Shade the decimal models to show each fraction.

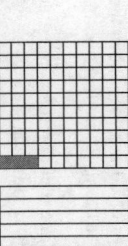

Part B: Compare the two decimals. Use >, <, or =. Explain.

$\frac{3}{10}$ $>$ $\frac{3}{100}$; 3 parts out of 10 is greater than 3 parts out of 100.

5. How can you use models to explain why 3.1 = 3.10?

Sample answer: If I used decimal grids, they would both have 3 full grids for the ones place, 1 column for the tens place, and no squares for the hundreds place.

NAME _____

DATE _____

SCORE _____

Countdown: 18 Weeks

1. Find the prime factorization for the number below.

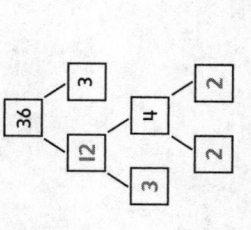

$$36 = \underline{\hspace{2cm}} \quad 2 \times 2 \times 3 \times 3$$

> **ONLINE TESTING**
> On the actual test, you might be asked to use a keyboard for entering the numbers in the boxes. In this book, you will be asked to write the numbers with a pencil instead.

2. Each box of paper clips contains 10^3 clips. The school store has 25 boxes.

Part A: What is the value of 10^3?

1,000

Part B: How many paper clips does the school store have?

25,000 paper clips

3. A recipe for pancakes calls for 3 cups of flour for every 2 tablespoons of sugar. Fill in the chart to find how many cups of flour are needed for 8 tablespoons of sugar.

Tablespoons of Sugar	Cups of Flour
2	3
4	6
6	9
8	12

4. Find 8 × 52 using an area model.

	50	2
8	8 × 50	8 × 2

$$8 \times 52 = (8 \times 50) + (8 \times 2)$$
$$= 400 + 16$$
$$= 416$$

5. Shade the box in front of the statements that are true.

☐ 3.240 > 3.24

▨ 2 and 34 hundredths = 2.34

▨ $2 + \frac{3}{10} = 2.3$

☐ 536 hundredths = 53.6

NAME _____ DATE _____

SCORE _____

Countdown: 17 Weeks

1. ABC Pens sells pens in boxes of 12. Their competitor XYZ Pens sells pens in boxes of 144. An office building is considering purchasing either 10^3 boxes from ABC Pens or 10^2 boxes from XYZ Pens.

   Part A: How many pens are in 10^3 boxes of ABC Pens?

   12,000 pens

   Part B: How many pens are in 10^2 boxes of XYZ Pens?

   14,400 pens

2. Circle the problems that have a correct solution.

   $$\begin{array}{r} 254 \\ \times\ 12 \\ \hline 3,048 \end{array}$$

   $$\begin{array}{r} 412 \\ \times\ 24 \\ \hline 9,788 \end{array}$$

   $$\begin{array}{r} 316 \\ \times\ 29 \\ \hline 9,164 \end{array}$$

   $$\begin{array}{r} 581 \\ \times\ 32 \\ \hline 18,592 \end{array}$$

   ONLINE TESTING
   On the actual test, you might be asked to click on the problem to put a circle around it. In this book, you will be asked to make the circles with a pencil instead.

3. The table below lists the number of students in each grade level of an elementary school. Estimate how many students are in the school by rounding. Show how you estimated.

Grade Level	Number of Students
Kindergarten	315
First Grade	378
Second Grade	412
Third Grade	351
Fourth Grade	401
Fifth Grade	345

   Sample answer: $(300 \times 2) + (400 \times 4) = 600 + 1,600 =$ 2,200 students

4. A penny is 1.52 mm thick. Write this number in expanded form.

   $$(1 \times 1) + \left(5 \times \frac{1}{10}\right) + \left(2 \times \frac{1}{100}\right)$$

5. The land area of Arizona is $(1 \times 100,000) + (1 \times 10,000) + (3 \times 1,000) + (9 \times 100) + (9 \times 10) + (8 \times 1)$ square miles.

   Part A: Write the correct digits in the boxes in order to put the number into standard form.

1	1	3 ,	9	9	8

   Part B: Write the area of Arizona in words.

   One hundred thirteen thousand, nine hundred, ninety-eight square miles

NAME

DATE

SCORE

Countdown: 16 Weeks

1. For the numbers 6, 7, and 42, circle the equations that are members of the fact family.

   (6 × 7 = 42)

   7 + 6 = 13

   7 × 42 = 6

   (42 ÷ 7 = 6)

   (7 × 6 = 42)

   (42 ÷ 6 = 7)

2. A candy company puts 200 pieces of candy inside the bag. In the month of July, the company sold 8,000,000 pieces of candy. Determine whether each statement will find the number of bags of candy the company sold in July.

   Yes No

   ☐ ☐ 8,000,000 ÷ 200

   ☐ ☐ 800,000 ÷ 200

   ☐ ☐ 8,000,000 ÷ 20

   ☐ ☐ 800,000 ÷ 20

   ☐ ☐ 80,000 ÷ 2

   ONLINE TESTING
   On the actual test, you might be asked to click on a box to shade it. In this book, you will be asked to shade the box with a pencil instead.

3. The table shows the amount that a painter charges for painting rooms. If your house has four bedrooms, two bathrooms, and three other rooms, how much will it cost to have the entire house painted?

Type of Room	Cost
Bedroom	$100
Bathroom	$50
Other Rooms	$120

   ($100 × 4) + ($50 × 2) + ($120 × 3) = $400 + $100 + $360 = $860

4. A roller coaster can take 24 riders in a single trip. 72 people went through the line to ride the roller coaster.

   Part A: How many trips did the roller coaster make?

   3 trips

   Part B: Write the multiplication and division fact family for this.

   3 × 24 = 72
   24 × 3 = 72
   72 ÷ 3 = 24
   72 ÷ 24 = 3

5. **Part A:** A school building has 67 classrooms in it. Four students have volunteered to clean the classrooms over summer break. Fill in the boxes to find how many rooms each student should clean.

   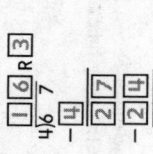

   Each student should clean ___16___ rooms.

   Part B: The building principal has offered to clean the left over rooms. How many rooms will she clean?

   3 rooms

Countdown

NAME _____ DATE _____

SCORE _____

Countdown: 15 Weeks

1. Part A: Eggs are sold by the dozen. If a chicken farm has produced 2,386 eggs, color in the box next to any expression that will estimate how many dozens can be packaged.

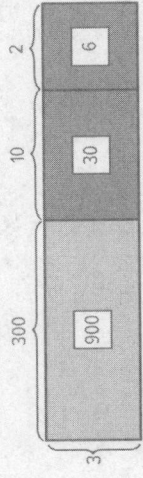

- ▨ $2{,}000 \div 12$
- ▨ $2{,}400 \div 12$
- ☐ $3{,}000 \div 12$
- ☐ $2{,}300 \div 12$

Part B: Estimate the number of dozens that can be packaged.

Sample answer: $2{,}400 \div 12 = 200$ dozens

2. Jerry and his two friends are going to bake cookies for a fundraiser. They need to bake 369 cookies in all. Use a model to find the number of cookies each person needs to bake.

Each person bakes ___123___ cookies.

> **ONLINE TESTING**
> On the actual test, you might be asked drag and drop groups of hundreds, tens, and ones to make the model. In this book, you will be asked to make the model by drawing instead.

3. Use the Distributive Property to draw a bar diagram and solve the problem.

$936 \div 3 =$ ___312___

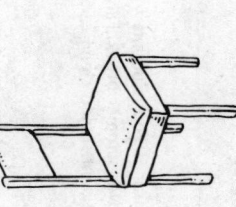

900	30	6
300	10	2

$300 + 10 + 2 = 312$

4. Chairs for a dining room set cost $78 each. Circle the equation that can be used to best estimate the cost of buying chairs for a family of 6.

$70 \times 6 = \$420$

$\$80 \times 6 = \$480$ ⟵ (circled)

$100 \times 6 = \$600$

$50 \times 6 = \$300$

5. Part A: Draw the decimal points on each number on the left side of the equation so that the difference is correct as shown.

$$1\,2\,3.4 - 6.2\,4 = 117.16$$

Part B: Check to make sure that the answer is reasonable by rounding.

$120 - 6 = 114$

NAME _____

DATE _____

SCORE _____

Countdown: 14 Weeks

1. There are 653 rubber bands in a desk drawer. The teacher wants to split them as evenly as possible among 63 students. Circle the equation that is the least accurate estimate for the number of rubber bands each student should receive.

$660 \div 66 = 10$

$650 \div 65 = 10$

$650 \div 50 = 13$

$480 \div 60 = 8$ (circled)

> **ONLINE TESTING**
> On the actual test, you might be asked click to draw a circle. In this book, you will be asked to draw the circle using a pencil.

2. Ms. Chen wants to purchase sets of Christmas lights to decorate her house. The lights cost $13 per package. She has saved $360 for the project.

Part A: How many packages of lights can she buy with $360?

27 packages

Part B: What is the remainder, and what does it represent?

$9; This is the amount Ms. Chen has left over.

Part C: Round to estimate the answer so you can check for reasonableness.

$360 ÷ $10 = 36

3. A farmer has a rectangular field to plow. The field has an area of 18,963 square yards. The field is shown below. Fill in the missing length.

147 yd

129 yd

4. Rayshawn is applying mulch along the fence in his backyard. For every 3 feet of length along the fence, he needs 2 bags of mulch. The fence is 126 feet long, and he has already finished 18 feet. How many more bags of mulch does he need?

Part A: Number each of the following steps to indicate the order in which they need to be completed to solve this problem.

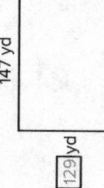

2	Divide by 3 to figure our how many more 3-foot segments there are.
1	Subtract 18 from 126 to find out how many more feet need landscaped.
3	Multiply by 2 to find the number of bags needed.

Part B: How many more bags of mulch are needed?

72 bags

5. Samuel went to the movies and purchased a ticket, a bag of popcorn, and a soda. He gave the cashier $20.00 and received $2.56 back in change. Fill in the cost of the popcorn in the table.

Item	Price
Ticket	$9.50
Popcorn	$5.32
Soda	$2.62

Countdown 13

NAME

DATE

SCORE

Countdown: 13 Weeks

1. Adrianna is making a photo collage for her parents' 20th anniversary party. Each poster board can fit 13 pictures, and she has 167 pictures.

Part A: How many poster boards can Adrianna fill?

[12 poster boards]

Part B: What is the remainder, and what does it represent?

II. This is how many pictures will be left over and not placed on a poster board.

2. Ahmal is trying to estimate how many boxes he will need to store his miniature car collection in. He has 538 cars, and 27 cars will fit nicely into the boxes he wants to buy.

Part A: Use the numbers below to choose the best pair that will estimate the number of boxes Ahmal needs, and write them in the blank spaces.

500   540   530   600   27   30   20

[540] ÷ [27]

ONLINE TESTING
On the actual test, you might be asked to drag and drop the numbers into the boxes. In this book, you will be asked to write the numbers using a pencil instead.

Part B: Estimate the number of boxes Ahmal will need.

[20 boxes]

Grade 5 • Countdown 13 Weeks   27

3. A construction company is looking at a rectangular piece of property on which to build an office building. The area of the property is 20,514 square yards. One side length is 78 yards. Draw the field and label the side lengths.

263 yards
78 yards

4. Compare $\frac{8}{10}$ and $\frac{8}{100}$.

Part A: Shade the decimal to match each fraction.

Part B: Fill in <, >, or =.

$\frac{8}{10}$ ◯ $\frac{8}{100}$

5. Complete the powers of 10 pattern in the top row of the table below. Then complete the pattern created in the bottom row by writing the corresponding power of 10 with an exponent.

610	6,100	61,000	610,000	6,100,000
61×10^1	61×10^2	61×10^3	61×10^4	61×10^5

28   Grade 5 • Countdown 13 Weeks

200

Grade 5 • Countdown 13 Weeks

Copyright © McGraw-Hill Education. Permission is granted to reproduce for classroom use.

NAME _____ DATE _____

SCORE _____

Countdown: 12 Weeks

1. Nick researched the weights of male and female lions. The chart shows his findings. Use rounding to estimate the difference in weight between a male and a female lion.

Male	Female
410.89 lbs	306.21 lbs

410 − 310 = 100 lbs.

2. A coach has timed a swimmer who completed two laps in the pool. The time for the swimmer's first lap was 57.12 seconds. The time for the swimmer's second lap was 61.8 seconds.

Part A: Shade the box under all correct ways of finding the swimmer's total time for both laps.

$$\begin{array}{r} 57.12 \\ + 61.8 \\ \hline \end{array} \qquad \begin{array}{r} 57.12 \\ + 61.80 \\ \hline \end{array} \qquad \begin{array}{r} 57.12 \\ + 61.8 \\ \hline \end{array}$$

■   ■   ☐

> **ONLINE TESTING**
> On the actual test, you might be asked click in order to shade the boxes. In this book, you will be asked to shade the boxes using a pencil instead.

Part B: Find the total time the swimmer took to swim both laps.

118.92 seconds

3. Antonio makes $13 for mowing his neighbors' lawns. He is saving up for a telescope that costs $49. Complete the table to help find out how many lawns Antonio will need to mow in order to make enough money to pay for the telescope.

1 Lawn	$13
2 Lawns	$26
3 Lawns	$39
4 Lawns	$52
5 Lawns	

4 _____ Lawns

4. The chart shows the total number of brownies sold at a bake sale on three different days. Place each of the numbers from the table in the blanks in a way that makes the addition problem the easiest. Explain your reasoning, and find the total.

Day	Brownies Sold
Monday	12
Tuesday	19
Wednesday	8

(12 + 8) + 19

12 and 8 are easy to add together, so they should be added first. The total is 39 brownies.

5. Jameson is trying to round 99.9999 to the nearest tenth.

Part A: Jameson asks three friends for the answer and gets three different responses. Circle the correct answer.

100.0 (circled)

99.0

99.9

Part B: Alana did the same problem but accidentally rounded to the nearest hundredth. She says she got the same answer. Is that possible? Explain.

It is possible. Because the digits are all nines, 99.9999 rounded to the nearest hundredth is also 100, but it should be written as 100.00.

Countdown

NAME

DATE

SCORE

Countdown: 11 Weeks

1. Write a real-world math problem that can be solved using the base-ten blocks below.

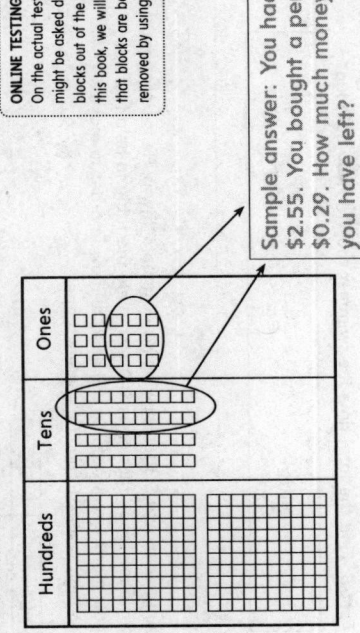

Hundreds	Tens	Ones

> **ONLINE TESTING**
> On the actual test, you might be asked drag the blocks out of the groups. In this book, we will show you that blocks are being removed by using arrows.

Sample answer: You had $2.55. You bought a pencil for $0.29. How much money do you have left?

2. Joshua is trying to subtract 8 − 4.13. He sets up the problem like this.

$$\begin{array}{r} 8 \\ -\ 4.13 \\ \hline \end{array}$$

Part A: What is wrong with Joshua's setup?

> The 8 is in the ones place and should be lined up with the 4.

Part B: What is the answer to Joshua's problem?

> 3.87

3. Look at the solution for doing 47 + 59 mentally. Select from the following properties to fill in the reasons for each step.

Commutative Property	Associative Property	Addition

$$47 + 59 = 47 + (56 + 3)$$
$$= 47 + (3 + 56) \quad \underline{\text{Commutative Property}}$$
$$= (47 + 3) + 56 \quad \underline{\text{Associative Property}}$$
$$= 50 + 56 \qquad\quad \underline{\text{Addition}}$$
$$= 106 \qquad\qquad\ \underline{\text{Addition}}$$

4. Janelle is asked to divide a number by 4. For each number in the Remainder column, determine whether the number is a possible remainder when dividing by 4. Shade either Yes or No. For any number that you marked as Yes, give an example of a division problem that has that number as a remainder when dividing by 4.

Remainder	Yes	No	Example
0	■	☐	16 ÷ 4
1	■	☐	17 ÷ 4
2	■	☐	18 ÷ 4
3	■	☐	19 ÷ 4
4	☐	■	
5	☐	■	

5. Circle the mistake in the prime factorization tree for 48.

48 → 4, 4 → 2, 2; 48 → (14) → 2, 2, 7

3. Javier sold 9 bags of cookies at $2.25 per bag. Molly sold 5 pieces of pie at $4.15 per piece.

Part A: How much did Javier earn?

$20.25

Part B: How much did Molly earn?

$20.75

Part C: Who earned more?

Molly

4. Jeremy bought a new computer. The length and width of the screen are shown. What is the area of the screen?

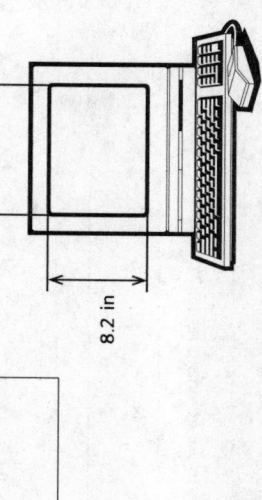

12.1 in

8.2 in

99.22 square inches

5. A local grocery stand sold $12,456.98 on a Friday. There were 289 customers. Circle the expression that would provide the best estimate for the average amount spent by a customer.

$13,000 ÷ 300

($12,000 ÷ 300)

$10,000 ÷ 300

$12,000 ÷ 200

NAME _____ DATE _____

SCORE _____

Countdown: 10 Weeks

1. An amusement park costs $47.50 admission for a day. A family of five wants to go to the park. Use the following set of numbers to fill in boxes that will help estimate the total cost for the family.

$40	$30	$50
5	1	10
$150	$250	$350

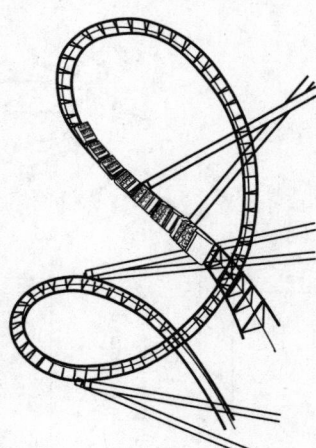

$50 × 5 = $250

2. Shade the models below to calculate 0.8 × 2.

0.8 × 2 = 1.6

NAME _____

DATE _____

SCORE _____

Countdown: 9 Weeks

1. Mr. Jackson took his new car on a family vacation. He drove the car 1,454.625 miles and used 4.5 tanks of gas. How many miles does he get on a single tank of gas?

[323.25 miles]

2. Sort the following multiplication problems into those that have answers that are greater than 1 and those that have answers that are less than 1.

0.89×10 0.012×10 0.034×10^2 1.29×10^3

Greater than 1	Less than 1
0.89×10	0.012×10
0.034×10^2	
1.29×10^3	

ONLINE TESTING
On the actual test, you might be asked drag the numbers into the groups. In this book, you will instead write the numbers using a pencil.

3. Max measured the length of a bug in science class to be 47.61 mm. Write this number in expanded form.

$(4 \times 10) + (7 \times 1) + \left(6 \times \frac{1}{10}\right) + \left(1 \times \frac{1}{100}\right)$

4. Jonathan is trying to calculate $(6.28 \times 50) \times 2$ without a calculator.

Part A: Reorganize the numbers to make the calculation easier.

$6.28 \times ($ 50 $\times$ 2 $)$

Part B: What property did you use to reorganize the numbers?

[The Associative Property of Multiplication.]

Part C: What is the answer to Jonathan's question?

[628]

5. Roland is trying to calculate $4.51 + 12.78$.

Part A: Explain what Roland is doing wrong in his setup?

```
  12.57
+  4.51
  5.767
```

[Roland has not lined up the decimal points.]

Part B: Find the correct solution to Roland's problem.

[17.29]

NAME

DATE

SCORE

Countdown: 8 Weeks

1. The bar diagram below can be represented with several expressions. However, not all of the ones below are correct. Circle any that are not correct, and evaluate each expression.

| 6 | 6 | 7 | 7 | 7 | 6 |

$6 + 6 + 7 + 7 + 7 + 6$ $6 \times 3 + 7 \times 3$

39   39

$3 \times (6 + 7)$ $3 \times (6 + 7) \times 3$ *(circled)*

39   117

2. Benjamin wants to find the area of a trapezoid-shaped garden. His teacher told him that the area can be found by first adding the lengths of the top and the bottom, then multiplying the sum by the height, and finally dividing the product by 2.

[trapezoid: 5 ft (top), 9 ft (bottom), height = 8 ft]

Part A: Shade in the box next to any expression that will find the area of the garden.

☑ $(5 + 9) \times 8 \div 2$ ☐ $[(5 + 9) \times 8] \div 2$

☐ $5 + 9 \times 8 \div 2$ ☐ $5 + 9 \times (8 \div 2)$

Part B: Evaluate the expression to find the area.

56 square feet

Grade 5 • Countdown 8 Weeks **37**

3. Both Andrea and Eileen are preparing to run a marathon, a 26.2 mile race. Each of the women begins her training by running 2 miles per day. Andrea says that she will double the amount that she runs per day with each passing week. Eileen says that she will add 5 miles to her daily run with each passing week. Use the two charts to determine who will be running *more than* 26.2 miles per day first.

Andrea

Week	Miles Per Day
1	2
2	4
3	8
4	16
5	32
6	64

Eileen

Week	Miles Per Day
1	2
2	7
3	12
4	17
5	22
6	27

Andrea will exceed 26.2 miles per day first.

4. A farmer is constructing a small fenced in area that can be describe with the ordered pairs (2, 3), (2, 8), (6, 8), and (6, 3). The units for both x and y are feet. Make a graph. Then find the amount of fencing he will need.

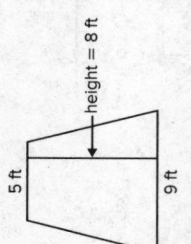

18 feet

5. A factory that produces piano keys churned out 545,952 keys in 12 months. A piano requires 88 keys. How many pianos can be produced using the keys from the first 3 months?

1,551 pianos

38 Grade 5 • Countdown 8 Weeks

NAME _____ **DATE** _____

SCORE _____

Countdown: 7 Weeks

1. Anya needs 7 cans of paint to put on three coats in her new living room.

Part A: How many cans of paint will it take to paint a single coat?

[7]

[3]

ONLINE TESTING
On the actual test, you might be asked drag the fraction onto the number line. In this book, you will draw the point using a pencil instead.

Part B: Place a point on the number line that represents the number.

1 2 3 4 5 6 7 8

2. 16 fiction books and 20 nonfiction books are to be put in giveaway bags. The number of fiction books in each bag will be the same as the number of nonfiction books in the bag.

Part A: What is the greatest number of bags that can be made?

4 bags

Part B: Jillian says that if the number of fiction books goes up to 18, then the number of bags that can be made will also go up. Is she correct? Why or why not?

No. Even though the number of fiction books gets larger, the new greatest common divisor of 18 and 20 is 2, so the number of bags gets smaller.

---

3. Mr. McDonald gave his math class the following problem.

18 pizzas need split among 12 families.

How many pizzas does each family get?

Different people in the class gave different answers. Circle the answers that are correct.

$\left(\dfrac{18}{12}\right)$ $\left(\dfrac{9}{6}\right)$ $\dfrac{1}{2}$ $\left(\dfrac{3}{2}\right)$ $\dfrac{9}{4}$ $\left(\dfrac{1}{2}\right)$ $\left(1\dfrac{3}{6}\right)$ $1\dfrac{1}{3}$

4. A local post office sells stamps in packs of 4, 6, and 7. Andy bought several packs of 4. Erin bought several packs of 6. Jarryn bought several packs of 7. Each of the three friends ended up with the same number of stamps. What is the smallest number of stamps that each person could have purchased?

84 stamps

5. Manuel has 13 granola bars to split among 5 people.

Part A: Express the number of granola bars that each person receives as a quotient and remainder. Interpret the result.

2 R 3. Each person gets 2 granola bars, and 3 are left over.

Part B: Express the number of granola bars that each person receives as a mixed number. Interpret the results.

$2\dfrac{3}{5}$. Each person receives 2 whole bars and $\dfrac{3}{5}$ of a bar. There are none left over.

NAME

DATE

Countdown: 6 Weeks

SCORE

1. The table shows Frances' quiz scores for the past grading period.

Part A: Fill in the fraction that Frances got correct for each quiz.

Number Incorrect	Number Correct	Fraction Correct
2	6	$\frac{6}{8}$
2	3	$\frac{3}{5}$
3	7	$\frac{7}{10}$
2	5	$\frac{5}{7}$
4	9	$\frac{9}{13}$

Part B: Frances' teacher has agreed to drop the worst test score for the grading period. Put her scores in order from least to greatest, and circle the score that can be dropped.

$\boxed{\frac{3}{5}}$, $\frac{9}{13}$, $\frac{7}{10}$, $\frac{5}{7}$, $\frac{6}{8}$

> **ONLINE TESTING**
> On the actual test, you might be asked to drag the numbers into their order. In this book, you will write the number using a pencil.

2. Compare each pair of numbers by using <, >, or =.

$\frac{14}{20}$ $(>)$ 0.65 $\frac{6}{25}$ $(=)$ 0.24 $\frac{1}{50}$ $(<)$ 0.2

3. Ian and Zion are trying to write $\frac{10}{25}$ as a decimal. Ian says that they should multiply the top and bottom by 4 and then convert to decimal form. Zion says that they should divide the top and bottom by 5, then multiply the top and bottom by 20, and then convert to decimal form. Who is correct? Show the work for each method and give the decimal answer.

> They are both correct. If Ian multiplies the top and bottom of $\frac{10}{25}$ by 4, the equivalent fraction is $\frac{40}{100}$. If Zion divides the top and bottom of $\frac{10}{25}$ by 5, he gets $\frac{2}{5}$. When he multiplies the top and bottom of $\frac{2}{5}$ by 20, he gets $\frac{40}{100}$. Both equal 0.40.

4. A runner wants to run 1,000 miles in one year. If he runs the same amount every day, use compatible numbers to estimate the number of miles he should run every day. Show your work.

> $900 \div 300 = 3$ miles per day

5. Jameson wants to construct a ladder that has 8 rungs. Each rung is 3.2 feet wide. The two sides measure 10.6 feet each. The wood is sold for $2.25 per linear foot. Find how much the wood to construct this ladder will cost Jameson.

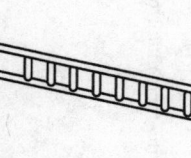

> $105.30

NAME

DATE

Countdown: 5 Weeks

SCORE

1. Kalim, Henry, and Joseph agreed to split the lawn mowing for the weekend. Kalim mowed $\frac{5}{12}$ of the lawn. Henry mowed $\frac{5}{12}$ of the lawn. Joseph mowed the rest. Fill in the chart with the fraction of the lawn that Joseph mowed, and put the fraction in lowest terms.

Kalim	$\frac{5}{12}$
Henry	$\frac{5}{12}$
Joseph	$\frac{1}{6}$

2. Circle the expression that is not equal to the others.

$\frac{2}{12} + \frac{1}{6}$ $\frac{1}{6} + \frac{1}{6}$

$\left(\frac{2}{12} + \frac{1}{2} + \frac{1}{3}\right)$

$\frac{2}{12} + \frac{1}{12} + \frac{1}{12}$

3. Theo opened a bag of marbles. $\frac{2}{15}$ of the marbles were red. $\frac{3}{5}$ of the marbles were blue.

Part A: What fraction of the marbles was red or blue?

$\frac{11}{15}$

Part B: What fraction of the marbles was neither red nor blue?

$\frac{4}{15}$

---

4. Victor claims that if two fractions are in lowest terms, then their sum will be in lowest terms as long as he uses the least common denominator. Drake is sure that he can find two fractions in lowest terms whose sum is not in lowest terms even if he uses the least common denominator.

Part A: Shade in the boxes next to the facts that Drake can use to prove Victor wrong.

☐ $\frac{2}{3} + \frac{1}{2}$

▓ $\frac{1}{4} + \frac{1}{12}$

▓ $\frac{1}{6} + \frac{1}{3}$

☐ $\frac{1}{18} + \frac{1}{3}$

Part B: For each box you shaded, add the fractions together using the least common denominator to show that the sum is not in lowest terms.

$\frac{1}{4} + \frac{1}{12} = \frac{3}{12} + \frac{1}{12} = \frac{4}{12}$, which is not in lowest terms.

$\frac{1}{6} + \frac{1}{3} = \frac{1}{6} + \frac{2}{6} = \frac{3}{6}$, which is not in lowest terms.

5. A company has purchased a large "L" shape plot of land on which to build a new factory. The coordinates of the "L" are (0, 0), (0,9), (4,9), (4,5), (10,5), and (10,0). The units are miles. Plot the "L" shape on the plane, and find the perimeter of the shape.

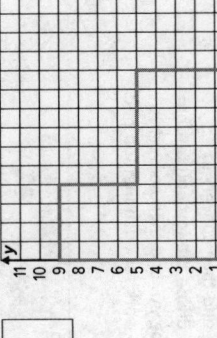

38 miles

NAME

DATE

SCORE

Countdown: 4 Weeks

1. Sydney planted $\frac{5}{9}$ of her fall flowers. She had 63 unplanted flowers to start.

Part A: How many flowers does she have left to plant?

28 flowers

Part B: How many more flowers would she need to plant in order to have planted $\frac{2}{3}$ of her flowers?

7 flowers

2. Place each of the following expressions into the two categories of "whole number" and "not a whole number" based on whether or not the product is a whole number.

2×7 $\frac{4}{5} \times 15$ $\frac{6}{7} \times 21$

$\frac{3}{4} \times 35$ $\frac{7}{11} \times 100$ $\frac{9}{13} \times 26$

Whole Number
$\frac{4}{5} \times 15$ $\frac{6}{7} \times 21$
$\frac{3}{4} \times 35$
$\frac{9}{13} \times 26$

Not a Whole Number
$\frac{2}{3} \times 7$ $\frac{3}{4} \times 35$
$\frac{7}{11} \times 100$

3. Winston has $\frac{1}{4}$ of a pound of chocolate to split equally among 4 friends.

Part A: How many pounds will each friend get?

$\frac{1}{16}$ pound

Part B: How many pounds will two of the friends get together? Write your answer in reduced terms.

$\frac{1}{8}$ pound

4. A playground is to be constructed in the shape of the rectangle shown.

$\frac{3}{5}$ mile

$\frac{5}{7}$ mile

Circle the correct expression for finding the area of the playground. Then find the area.

$\frac{3}{5} + \frac{5}{7}$ $\frac{3}{5} - \frac{5}{7}$ $\frac{3}{5} \div \frac{5}{7}$

$\frac{3}{5} \times \frac{5}{7}$

Area = $\frac{3}{7}$ _____ square miles

5. Ms. Trenton measured the rainfall for five consecutive days. Place the days in order from least to greatest amount of rainfall.

Monday	1.01 inches
Tuesday	1.001 inches
Wednesday	1.101 inches
Thursday	1.10 inches
Friday	1.11 inches

Tuesday, Monday, Thursday, Wednesday, Friday

Countdown

Countdown: 3 Weeks

1. Two teams of scientists measured the length of a cactus needles for a study on desert plant growth. The first team measured the length to the nearest quarter inch and reported a length of $5\frac{3}{4}$ inches. The second team measured the length to the nearest eighth inch and reported a length of $5\frac{7}{8}$ inches. Shade the box under "Yes" or "No" if the length given could be the actual length of the cactus needle.

ONLINE TESTING
On the actual test, you might be asked to shade the boxes by clicking on them. In this book, you will instead shade the boxes by using a pencil.

Yes	No	
☐	☐	5.876 inches
☐	☐	5.85 inches
☐	☐	5.741 inches
☐	☐	5.83 inches

2. In nautical uses, a fathom is a length of measure that is equivalent to 6 feet. Fill in the following conversion chart for fathoms.

880 fathoms = ___ mile

1 fathom = 2 yards

1 fathom = 72 inches

3. Jayne, Carlos, June, and Pedro each measured their dogs' weights, but each used a different measurement. Place the four dogs in order from lightest to heaviest.

Jayne's Dog	42 lbs
Carlos' Dog	624 oz
June's Dog	41 lbs, 7 oz
Pedro's Dog	665 oz

Carlos' Dog, June's Dog, Pedro's Dog, Jayne's Dog

4. A certain species of seaweed doubles in weight every week.

Part A: Fill in the table with the weight for each of the first four weeks. Write your answers as a combination of pounds and ounces.

Week 1	1 pound 5 ounces
Week 2	2 pounds 10 ounces
Week 3	5 pounds 4 ounces
Week 4	10 pounds 8 ounces

Part B: Julian claims that the weight for Week 4 can be written as 10.5 pounds. Is he correct?

Yes. 10.5 pounds is the same as 10 pounds 8 ounces.

5. Bernardo times his drive to and from work. The drive in to work on Monday took 47.23 minutes. The drive home took 56.2 minutes. Circle any correct way of setting up the total time Bernardo spent in the car, and find the answer.

47.23 + 56.2   47.23 + 56.20   47.23 + 56.2

Total = 103.43 minutes

NAME

DATE

SCORE

Countdown: 2 Weeks

ONLINE TESTING
On the actual test, you might be asked to drag each symbol into its circle. In this book, you will write the symbol by using a pencil instead.

1. Fill in <, >, or = to make each of the following statements true.

16 cups $=$ 8 pints

19 quarts $<$ 5 gallons

81 cups $>$ 20 quarts

95 cups $<$ 6 gallons

2. Joy started with 2.75 gallons of milk. She used 1.5 pints to make mashed potatoes and another cup to make cookies. How much milks does Joy have left? Give your answers three different ways.

40 _____ cups

20 _____ pints

2.5 _____ gallons

3. There are approximately 3.1 miles in 5 kilometers. Thaddeus is supposed to ride his bike for 20 kilometers for a charity ride.

Part A: How many meters is this ride?

20,000 meters

Part B: How many miles is this ride?

12.4 miles

4. Place each of the following expressions into the two categories of "Greater than 1" and "Less than 1" based on the value of the product.

$$\frac{2}{7} \times 4 \qquad \frac{3}{19} \times 5 \qquad \frac{12}{33} \times 3$$

$$\frac{10}{21} \times 2 \qquad \frac{33}{100} \times 3 \qquad \frac{16}{75} \times 5$$

Greater than 1

$$\frac{2}{7} \times 4 \qquad \frac{12}{33} \times 3$$

$$\frac{16}{75} \times 5$$

Less than 1

$$\frac{3}{19} \times 5 \qquad \frac{10}{21} \times 2$$

$$\frac{33}{100} \times 3$$

5. The table below lists the number of pumpkins sold at a pumpkin farm during the course of one week. Estimate how many pumpkins were sold in total for the week. Show how you estimated.

Day	Number of Pumpkins
Monday	212
Tuesday	198
Wednesday	276
Thursday	181
Friday	303
Saturday	315

$$(200 \times 3) + (300 \times 3) = 600 + 900$$
$$= 1,500 \text{ pumpkins}$$

Countdown

NAME

DATE

SCORE

Countdown: 1 Week

1. One of the figures is an obtuse isosceles triangle. Circle the obtuse isosceles triangle.

2. An architect is asked to describe the shape of a floor plan for a kitchen, which is shown below.

Part A: Write all of the accurate names for the shape of the room in the given space.

Parallelogram   Rectangle
Rhombus   Quadrilateral
Trapezoid   Square

Parallelogram, Rectangle, Quadrilateral

Part B: Which name is the most appropriate for the shape of the room?

Rectangle

3. Shane pitches a tent for his weekend camping trip.

Part A: Circle any of the following shapes that are faces of the tent.

Triangle   Rectangle   Pentagon   Square   Hexagon

Part B: What is an appropriate name for the shape of the tent?

Triangular Prism

4. Uma is building the following bookshelf. What is the volume of the bookshelf?

3.1 m
0.4 m
1.2 m
2.3 m
1.6 m

$V = $ _____ 2.192 cubic meters

5. A law firm hires the same number of lawyers every year. At the end of 12 years, the firm has hired 48 lawyers. How many lawyers has the firm hired in the last 5 years?

20 lawyers

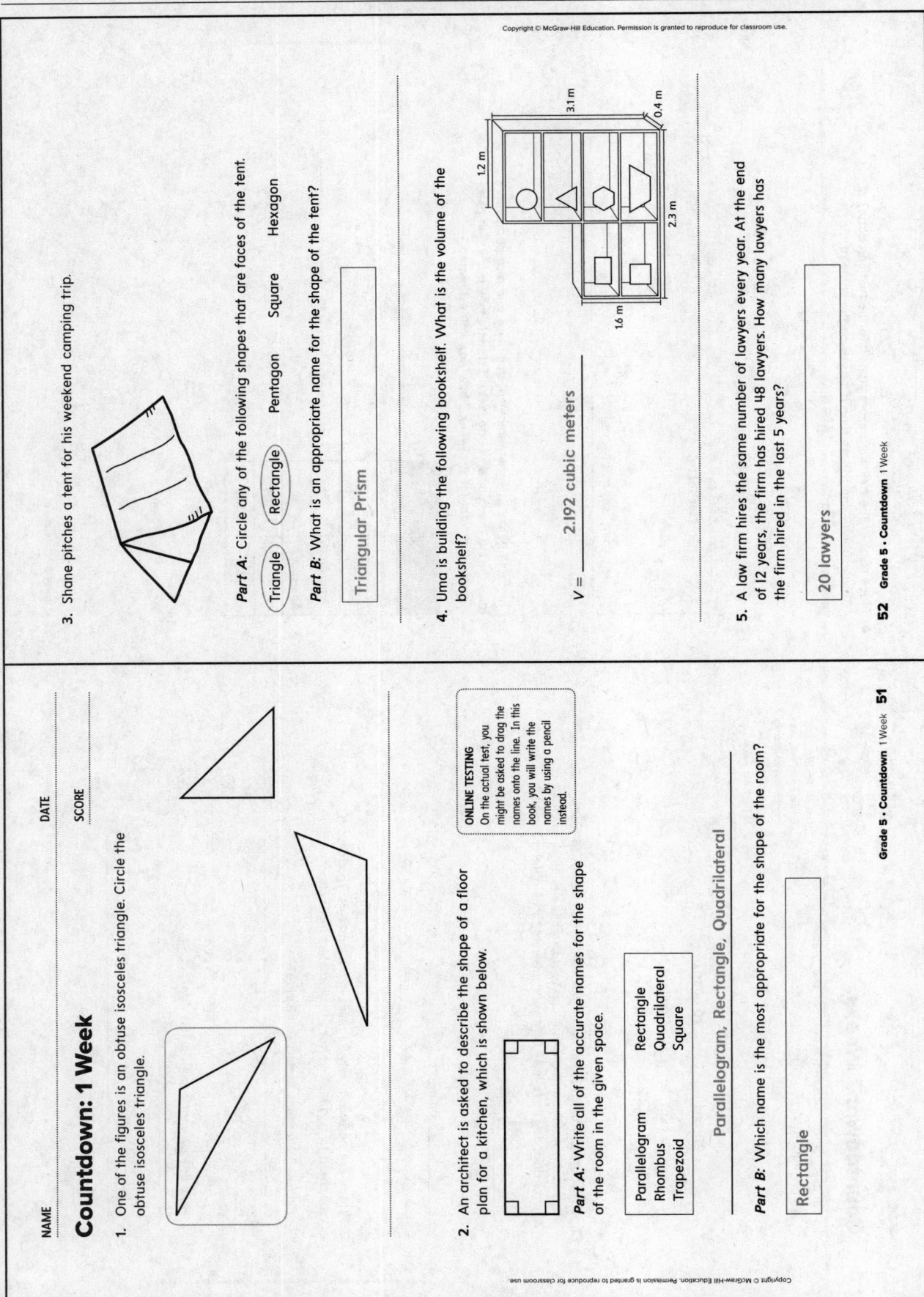

NAME

DATE

Chapter 1 Test

SCORE

1. A smart phone company sold 17,468,164 smart phones last year.

Part A: Fill in the place value chart for the number of smart phones sold by the company.

Millions			Thousands			Ones		
hundreds	tens	ones	hundreds	tens	ones	hundreds	tens	ones
	1	7	4	6	8	1	6	4

10,000,000
7,000,000
400,000
60,000
8,000
100
60
4

Part B: Write the number in words.

> Seventeen million, four hundred sixty-eight thousand, one hundred sixty-four

Part C: Write the expanded form of the number.

> $1 \times 10,000,000 + 7 \times 1,000,000 + 4 \times 100,000 + 6 \times 10,000 + 8 \times 1,000 + 1 \times 100 + 6 \times 10 + 4$

2. Alejandro is asked by his teacher to write the smallest five-digit number he can using the digits 1, 3, 5, 7, and 9.

Part A: If Alejandro is only allowed to use each digit once, what is the smallest five-digit number he can write?

> 13,579

Part B: If Alejandro is allowed to use each digit more than once, what is the smallest five-digit number he can write?

> 11,111

---

3. A student measures the length of a postage stamp to be 0.34 inches. He writes down the length as $\frac{34}{1000}$ inches. What is this student's mistake?

> The length of the stamp is $\frac{34}{100}$. The student has an extra 0 in the denominator.

4. The table below shows the attendance at a college's first four football games of the season. Put the numbers in order from least to greatest. Is attendance getting smaller or larger?

Date	Attendance
September 28	92,112
September 21	90,912
September 14	88,001
September 7	87,314

> 87,314; 88,001; 90,912; 92,112.
>
> The attendance is getting larger.

5. Shade in the following pictures to show the fractions for 0.3 and 0.30. What can you say about these two numbers by looking at the pictures?

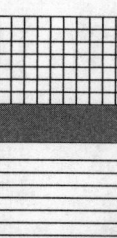

> The numbers are equal because the same area is shaded.

6. Jada's father sent her into the hardware store to find a bolt that is 0.625 inches long. Jada sees the following measurements for bolts. Circle the one is she supposed to buy.

$\left(\dfrac{625}{1,000}\right)$ $\dfrac{625}{100}$ $\dfrac{625}{10,000}$

7. The table below shows decimals and fractions. Fill in the table so that the left column has equals values as the right column.

0.234	$\dfrac{234}{1,000}$
0.0015	$\dfrac{15}{10,000}$
0.062	$\dfrac{62}{1,000}$
0.06	$\dfrac{6}{100}$

8. Paul is weighing a plant for a science project. The weight of the plant is 0.777 kg.

Part A: The value of the digit in the tenths place is how many times the value of the digit in the hundredths place? _____ 10

Part B: The value of the digit in the hundredths place is how many times the value of the digit in the thousandths place? _____ 10

Part C: The value of the digit in the tenths place is how many times the value of the digit in the thousandths place? _____ 100

9. A new player's batting average for the year is 0.289. Write this number out in expanded form.

$$2 \times \frac{1}{10} + 8 \times \frac{1}{100} + 9 \times \frac{1}{1,000}$$

10. Which of the following is not equal to the others? Circle the answer.

4.81

Four and eighty-one hundredths

$\left(4 \times 1 + 8 \times \dfrac{1}{100} + 1 \times \dfrac{1}{1,000}\right)$

11. The following chart lists the height of six children from a family. Place the heights in order from greatest to least.

4.25 feet	3.51 feet	3.49 feet
4.2 feet	4.56 feet	3.15 feet

4.56, 4.25, 4.2, 3.51, 3.49, 3.15

12. Sharon writes the weights of her marbles in order from least to greatest, but she makes a mistake. Circle the two numbers that must be switched so that all of the numbers are in the correct order.

1.022 g   1.02 g   1.2 g   1.202 g   1.22 g

13. Mrs. Shen had some eggs in her refrigerator. She bought a pack of twelve eggs for baking. She used six of the eggs and now has nine left in her refrigerator. How many eggs did Mrs. Shen have in her refrigerator before she bought more?

3 eggs

14. The Suarez family takes three days to drive to their vacation in North Carolina. The chart shows how many miles the family drove each day. If the family drove 31 less miles on Sunday than they did on Saturday and the total trip was 823 miles, fill in the missing values on the chart.

Day	Miles
Friday	352
Saturday	251
Sunday	220

15. Janice went out to eat and bought a hamburger, a bag of chips, and a drink. The hamburger cost $2.57, and the chips cost $1.25. Janice gave the cashier $20.00 and received $14.39 in change. How much did the drink cost?

$1.79

16. A student is struggling to understand the difference between 0.77 and 0.077.

Part A: Explain why 0.77 > 0.077

0.77 has a seven in the tenths place, and 0.077 has a 0 in the tenth place, so 0.77 is greater than 0.077.

Part B: Put 0.77, 0.077, and 0.707 in order from least to greatest.

0.077, 0.707, 0.77

17. ***Part A:*** Zoe says that "one hundred one thousand" is the same as "one thousand one hundred." Why is she incorrect?

"One hundred one thousand" is 101,000. "One thousand one hundred" is 1,100. These are different numbers.

Part B: Zoe also thinks that "one hundred one thousandths" is the same as "one hundred and one thousandths." Why is she incorrect?

"One hundred one thousandths" is 0.101. "One hundred and one hundred" is 100.001. These are different numbers.

18. Mrs. Hodge has asked her class to use the digits 3, 9, 6, 6, 2, 1 to make a number that is in between 310,000 and 330,000. Four students came up with the following answers. Shade the box next to the answers that are correct.

- ▣ 319,626
- ▣ 316,269
- ▣ 321,669
- ☐ 328,169

19. Place a decimal point in the following number so that the number is between 34 and 35.

3 4 . 3 4 3

20. The local news station found out that 123,000 people moved out of the city last year. Shade the box next to the correct way the news reporter should read this number during her report.

- ▣ One hundred twenty-three thousand
- ☐ One hundred and twenty three thousand
- ☐ One hundred twenty three thousandths
- ☐ One hundred and twenty-three thousandths

Chapter Tests

NAME

DATE

SCORE

Chapter 2 Test

1. Jed is buying water bottles for his soccer team. Because all of the packages of water bottles cost about the same price, Jed decides to buy the package of water bottles that provides the greatest total ounces.

   Part A: Complete the table below with the number of ounces per package.

Water Bottle Packages	
Water Bottles in a Package	Ounces per bottle
1	128
12	12
24	8
32	6

Water Bottles in a Package	1	12	24	32
Total Ounces in a Package	128	144	192	192

   Part B: Which package provides the greatest total ounces of water? Justify your response.

   24-bottle or 32-bottle package; both of these packages provide 192 total ounces of liquid.

2. Teams of 4, 5, or 6 members are permitted in a competition. If the grand prize will be divided in whole dollar amounts, evenly among the members of the winning team, which of the grand prizes is possible for this competition?

	Yes	No
$120		
$90		
$48		
$480		

3. Adrianna has 30 bills in her wallet. Some are $1 bills, some are $10 bills, and some are $100 bills. Which of the possible combination of bills in Andrea's wallet has the greatest value? Explain how you solved the problem.

Possible Bill Combinations		
$1 Bills	$10 Bills	$100 Bills
15	15	0
12	17	1
28	0	2
4	26	0

   If Adrianna had 12 ones, 17 tens, and 1 hundred, she would have the largest amount of $282. I multiplied each column by the appropriate power of 10 and then added across the rows.

4. The table shows the ticket cost of certain prizes at a fair.

Prizes	Tickets Needed
Stuffed Animals	125
Noise Maker	64
Sticky Hand	38
Pencil	15

   Which combination of prizes can you buy if you earned 432 tickets?

Prize Combinations	Yes	No
7 Noise Makers		
3 Stuffed Animals, 2 Pencils		
6 Sticky Hands, 15 Pencils		
2 Stuffed Animal, 4 Sticky Hands		

5. Over the period of one month 159 dogs visited the dog park. Suppose the same number of dogs visited each month for 1 year. How is this total different from the year before when 95 dogs visited the dog park every 3 months for the year? Show your work.

   $159 \times 12 = 1{,}908$ dogs this year
   95 dogs $\times 4 = 380$ dogs the year before
   $1{,}908 - 380 = 1{,}528$ more dogs this year

6. Part A: Complete the powers of 10 pattern in the top row of the table below. Then complete the pattern created in the bottom row by writing the corresponding power of 10 with an exponent.

780	7,800	78,000	780,000	7,800,000
78×10^1	78×10^2	78×10^3	78×10^4	78×10^5

Part B: Analyze each pattern. Explain the relationship between the top row pattern and the bottom row pattern. What does this pattern mean when considering the numbers above?

Sample answer: The number of zeros in the number in the top row is the same as the exponent following the 10 in the bottom row. It means that each extra power of 10 increases the value of the number 78 by ten times.

7. The following clues are given about a pail of marbles.

1.	There are between 700 and 800 marbles in the pail.
2.	The marbles were purchased in 8 equally-sized bags.
3.	The product of all the digits is 70.

How many marbles are in the pail? Explain how you figured it out.

752 marbles. Sample answer: The product of all the digits, 70, has a prime factorization of $2 \times 5 \times 7$. I knew the digits in the number of marbles had to be some combination of 2, 5, and 7. Since the number of marbles was between 700 and 800, I knew the first digit had to be a 7. The options were 725 or 752. Only 752 has a factor of 8.

8. A class will purchase 24 tickets to a play. Each ticket costs $78. Use an area model to find the total cost for the tickets.

Part A: Write an equation to represent the use of partial products to complete each part of the area model.

	70	8
20	$20 \times 70 = 1400$	$20 \times 8 = 160$
4	$4 \times 70 = 280$	$4 \times 8 = 32$

Part B: What is the total cost of the tickets?

$1872

9. To the right is an example of Jordan's work on a recent test.

$$
\begin{array}{r}
\overset{\scriptstyle 1\ 6}{117} \\
\times\ 19 \\
\hline
1053 \\
+\ 117 \\
\hline
1170
\end{array}
$$

Part A: Identify Jordan's error.

Sample answer: Jordan forgot that he was multiplying tens in the second row, so 117 meant 117 tens, or 1170, for a total of 2223.

Part B: Explain how if Jordan estimated the product he would have seen that his answer was not reasonable?

Sample answer: $120 \times 20 = 2400$, so Jordan's answer is too low.

10. A scientist is labeling insects for his collection. He knows the approximate weights of different amounts of each insect. Use the table to complete the weights shown.

Weights	
1000 Ants	4 grams
1000 Centipedes	140 grams
1 Spider	1 gram
100 Honey bees	1 gram

10,000 Ants _____ 40 _____ grams

100 Centipedes _____ 14 _____ grams

100 Spiders _____ 100 _____ grams

1000 Honey bees _____ 10 _____ grams

Chapter Tests

11. A company makes straws. The table shows the number of straws that are packaged in their different-sized boxes each hour.

Number of Straws in Each Box	Number of Boxes	Total Straws
10^2	95	9,500
10^2	55	5,500
10^4	185	1,850,000
10^3	115	115,000

Part A: Complete the table.

Part B: How would you write the first column of numbers as repeated multiplication expressions?

Sample answer: 10 × 10; 10 × 10; 10 × 10 × 10 × 10; 10 × 10 × 10

12. The stairway shown is made by putting 10 cement blocks together. If each cement block costs $23, how much would 10 complete stairways cost? Explain.

1 stairway is 10 cement blocks × $23 = $230. 10 stairways × $230 = $2,300.

13. A physician recorded a person's resting heart rate to be 87 beats per minute. Complete the table to estimate the total number of times the person's heart would beat for each interval shown.

Number of Minutes	1	10	100	1000
Number of Heartbeats	87	870	8,700	87,000

14. The product of 54 and another number is 8,720. Use the table to help you estimate the range for the other number.

Multiplication	Product
54 × 10	540
54 × 100	5,400
54 × 150	8,100
54 × 175	9,450
54 × 1,000	54,000

Sample answer: The number is between 150 and 175.

15. Rent costs $478 each month. Complete the partial product diagram for how much rent costs over a year.

	400	70	8
10	400 × 10 = 4000	70 × 10 = 700	8 × 10 = 80
2	400 × 2 = 800	70 × 2 = 140	8 × 2 = 16

The rent costs $5,736 for a year.

16. Skateboarders count rotations in half-turns of 180 degrees.

Part A: If the rotation record is 4 half-turns, how many total degrees is the record? **Part A:** 720 degrees

Part B: If Sean performed the rotation record 4 times, how many total degrees did he turn? **Part B:** 2,880 degrees

Part C: Explain how Sean's performance compares to a single half-turn. **Part C:** Sample answer: Sean performed four half-turns four times (4 × 4), so Sean's performance was 16 times a single half-turn.

17. A kilogram is 10^3 grams.

Part A: Write 10^3 grams in expanded notation. 10 × 10 × 10 grams

Part B: Suppose a package weighs 2 kilograms. How many grams is it? Explain. 2 × 10 × 10 × 10 = 2,000 grams

NAME _____ DATE _____

SCORE _____

Chapter 3 Test

1. Circle the fact that does not belong to the multiplication fact family.

   $3 \times 9 = 27$ $27 \div 3 = 9$

   (3 × 3 = 9) $27 \div 9 = 3$

2. A group of 36 cans of juice is divided among four children.

   Part A: If each child receives c cans, write an equation to find the unknown.

   [$36 \div 4 = c$]

   Part B: Find the unknown value c.

   [c = 9 cans]

3. Write and solve a division problem that is modeled by the picture.

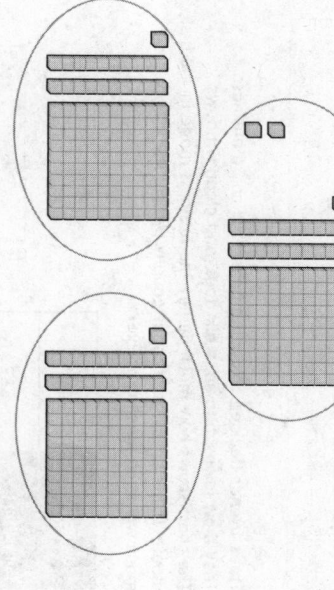

   [$365 \div 3 = 121$ R 2]

4. Bradford is taking down bulbs from a holiday decoration. The bulbs are put into boxes that can hold 6. He has 81 bulbs. What is the remainder? What is the meaning of the remainder?

   [The remainder is 3 bulbs. Sample answer: The last box of bulbs will only be half full.]

5. Circle the mistake in the division problem.

   $$\begin{array}{r} 25R2 \\ 4\overline{)92} \\ -8 \\ \hline (22) \\ -20 \\ \hline 2 \end{array}$$

6. Daiki is trying to sell 40 cupcakes that she made for a bake sale. He would like to sell them in boxes of 6. How many are left over that will need to be sold individually?

   [4 cupcakes]

7. **Part A:** Fill in the chart with solutions to the division problems.

9,000 ÷ 3	9,000 ÷ 30	9,000 ÷ 300	9,000 ÷ 3,000
3,000	300	30	3

   Part B: Describe the pattern.

   [As the divisor is multiplied by 10, the quotient is divided by 10.]

Chapter Tests

8. Malik says that 16,000 ÷ 4,000 is the same as 160 ÷ 40. Is Malik correct? Explain.

> Malik is correct. If we cross out three zeros from both numbers is the first problem, we get 16 ÷ 4. If we cross out one zero from both numbers in the second problem, we also get 16 ÷ 4. In either case, the answer is 4.

9. Three friends decided to open a household chore business. They mow lawns, babysit, walk dogs, and clean windows. The chart shows how much money the business made in the first month. If the earnings are split equally among the friends, how much will each person receive?

Mowing Lawns	$67
Babysitting	$82
Walking Dogs	$17
Cleaning Windows	$26

> $64

10. Madison and her friend Gabriel are both trying to estimate 182 ÷ 91. They both round to different place values.

Madison	Gabriel
180 ÷ 90	200 ÷ 100

Explain each student's thinking. What are their estimates? Are both correct?

> Madison rounded to the nearest ten. Gabriel rounded to the nearest 100. Both estimates are 2. Both are correct.

11. Use base ten blocks to model and solve the division problem 246 ÷ 2.

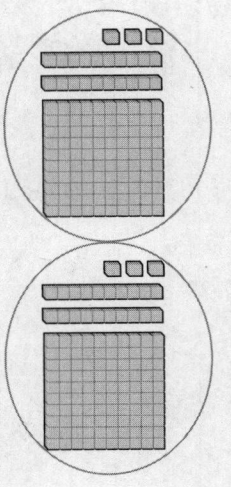

246 ÷ 2 = ___123___

12. A bookshelf has 5 shelves on it. There are 155 books that need put away.

Part A: Use the distributive property and the picture below to find how many books should be on each shelf.

	20	10	1
5	100	50	5

20 + 10 + 1 = 31 books

Part B: Sonny did the problem with a different picture.

	10	10	10	1
5	50	50	50	5

Is he correct? Explain.

> Yes, Sonny is correct. 155 can be split into 100 + 50 + 5 or into 50 + 50 + 50 + 5. The answer is still 31 books.

13. There are 144 roses that need put onto 8 tables at a wedding reception. How many roses should be put on each table?

| 18 roses |

14. Nine friends want to go to an amusement park. The total bill for all nine tickets is $423.

Part A: How much is each ticket?

| $47 |

Part B: Estimate to check your answer.

| Sample answer: $400 ÷ 10 = $40. The answer is reasonable. |

15. Daksha was asked by her teacher to predict the number of digits in the following quotients without actually dividing. How can she do this?

Part A: 834 ÷ 7

| 7 goes into 8 one time. Because 834 has three digits, the quotient will also have three digits. |

Part B: 567 ÷ 6

| 6 does not go into 5. Because 567 has three digits, the quotient will have 2 digits. |

16. Describe the student's error in the following division problem, and do the problem correctly.

$$\begin{array}{r} 29 \\ 3\overline{)627} \\ -6 \\ \hline 27 \\ -27 \\ \hline 0 \end{array}$$

$$\begin{array}{r} 209 \\ 3\overline{)627} \\ -6 \\ \hline 27 \\ -27 \\ \hline 0 \end{array}$$

| The student left out the 0 in the quotient. 3 goes into 2 zero times. |

17. Lamar runs 8 miles a day. He wants to know how many miles he ran in a particular month. Is there too much information or not enough information to solve this problem? Shade the box next to the correct description. If there is too much information, name the extra information and solve the problem. If there is not enough information, describe what Lamar would need to know to solve the problem.

☐ Too much information   ▨ Not enough information

| He needs to know how many days are in the month. |

18. *Part A:* A table at a party seats 8 guests. There are 71 guests expected. Find the number of tables needed and interpret the remainder.

| The remainder is 7 guests. Since 71 ÷ 8 = 8 R 7, there are 8 full tables and an extra table with the remainder. So 9 tables are needed. |

Part B: Christian is making birdhouses. Each birdhouse requires 10 screws. Christian has 81 screws. Find the number of birdhouses Christian can make and interpret the remainder.

| The remainder is 1 screw. 81 ÷ 10 = 8 R 1. Christian can build 8 birdhouses and will have 1 screw left over. |

NAME

DATE

SCORE

Chapter 4 Test

1. Circle any that would not be good ways of estimating 328 ÷ 32.

330 ÷ 30 = 11

400 ÷ 20 = 20 ⟵(circled)

300 ÷ 30 = 10

300 ÷ 20 = 15 ⟵(circled)

2. A scientist is studying fish population in an area of the ocean that measures 18,314 square miles. He wants to divide the area into 87 equal size portions to make the study more manageable. Estimate how large each area will be. Show your work.

Sample answer: 18,000 ÷ 90 = 200 square miles

3. A cab company is interested in how many vehicles it should station outside a particular hotel. This company has vans that hold 14 people. By looking at checkout patterns, the company determined that 160 people leave this hotel per day for the airport. Draw a model with base ten blocks to figure out how many vans the company should have ready in order to seat 160 guests.

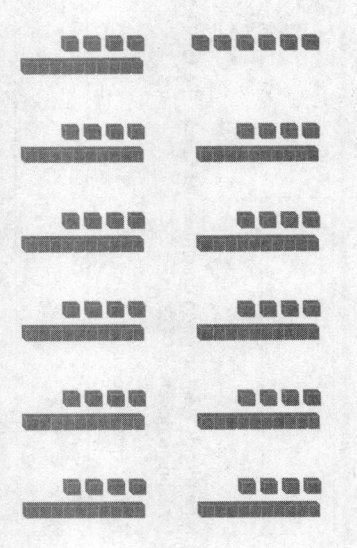

12 ____ vans

4. Mrs. Canzales needs to buy gallons of paint to paint her new house. Each can of paint cost $18. She has $310.

Part A: How many gallons of paint can she buy?

17 gallons

Part B: What is the remainder, and what does it mean?

$4. This is the amount of money she has left over.

Part C: Estimate to check your answer. Show your work.

300 ÷ 20 = 15 gallons. The answer is reasonable.

5. Quentin is asked by his teacher to write a division problem with quotient of 23 and a remainder of 3. Quentin wrote the following problem: 347 ÷ 15.

Part A: Complete Quentin's problem to show that it is not correct.

347 ÷ 15 = 23 R 2. The quotient is correct, but the remainder is not.

Part B: Help Quentin fix his problem by changing the 347 by a small amount.

348 ÷ 15 = 23 R 3

6. A factory produces 12,376 granola bars in 52 minutes. How many granola bars does the factory produce per minute?

238 granola bars per minute

7. A school recently received a donation for $7,072. The school has 17 different student organizations and wants to split the gift evenly. Miles has been asked to help figure out how much each organization should get. He starts the problem off like this:

$$\begin{array}{r} 3 \\ 17\overline{)7,072} \\ -51 \\ \hline 19 \end{array}$$

At this point, Miles knows that the 3 is not correct because 19 is bigger than 17. He adjusts the 3 to a 2 and gets

$$\begin{array}{r} 2 \\ 17\overline{)7,072} \\ -34 \\ \hline 36 \end{array}$$

Miles is now confused. What did he do wrong?

> Miles adjusted the quotient in the wrong direction. When the 3 didn't work, he should have tried 4, not 2.

8. Carlos' goal is to keep track of the total amount that he has run. After 16 months, his total is 2,400 miles. His coach, however, wants to know how much he ran in the past year. If Carlos ran the same amount every month, find his total distance for the past year.

> 1,800 miles

9. *Part A:* Fill in the following table with quotients and remainders.

Division Problem	Quotient	Remainder
6245 ÷ 12	520	5
6246 ÷ 12	520	6
6247 ÷ 12	520	7

Part B: What pattern do you notice?

> As the dividend goes up by 1, the remainder goes up by 1.

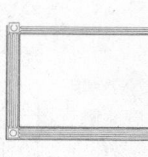
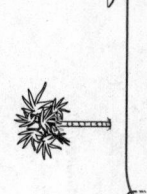

10. The college marching band raised $15,708 to help pay for a trip to a national parade. There are 132 students in the band. How much will each student receive in order to help pay for his or her airfare? Will there be any left over?

> $119. There is no remainder, so there is no money left over.

11. An art gallery purchases the same amount of prints per year to sell in its gift shop. In the last 7 years, the gallery has purchased a total of 882 prints. How many did the gallery purchase in its first 3 years?

> 378 prints

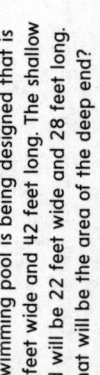

12. A swimming pool is being designed that is 22 feet wide and 42 feet long. The shallow end will be 22 feet wide and 28 feet long. What will be the area of the deep end?

> 308 square feet

13. Tyron is asked to find the missing value *h* in the equation:
$12,336 \div h = 16$
His friend Seamus says that he can rewrite this using another member of the same fact family and then solve the problem.

Part A: Fill in the boxes to rewrite the equation using another member of the fact family.

$$12,336 \div \boxed{} = 16 \qquad \boxed{} = h$$

Part B: Find the missing value *h*.

$h = \boxed{771}$

Copyright © McGraw-Hill Education. Permission is granted to reproduce for classroom use.

Chapter Tests

14. A restaurant sells chicken in packs of 6 pieces. The restaurant orders a large bag of chicken and splits the pieces into packs of 6, but there are some pieces left over. Shade the boxes next to any number that is a possible remainder, then explain your reasoning.

▓ 0   □ 3   □ 6

▓ 1   □ 4   □ 7

▓ 2   □ 5   □ 8

> When dividing by 6, the only remainders possible are numbers that are less than 6.

15. ABC Electronics produces a circuit board that can be used in computers. ABC's factory produced 18,270 boards last week. ABC supplies these boards to 90 different computer manufacturers and wants to give an equal amount to each manufacturer.

Part A: Fill in the division fact with compatible numbers to estimate how many boards each manufacturer should receive.

18,000 ÷ 90 = 200 boards

Part B: Find the exact number of boards each manufacturer will receive.

203 boards

Part C: Is your estimate greater than or less than the actual number? Explain how you could have known this ahead of time.

> The estimate is less than the actual number. I could have known this would be true because I rounded 18,270 down to 18,000 for the estimate.

16. 2,134 ÷ 8 has a remainder of 6. Circle all of the following facts that also have remainders of 6.

(2,126 ÷ 8)   2,127 ÷ 8   2,128 ÷ 8

(2,142 ÷ 8)   2,143 ÷ 8   2,144 ÷ 8

17. Hyun usually runs races that are 10 kilometers long. His time for a 10 kilometer race is 50 minutes. There is a charity race this weekend that is 9 kilometers long. What should Hyun expect his time to be at the race this weekend?

45 minutes

18. There is a set of swings to split among 17 different playgrounds in a city. Each playground gets 4 swings, and there are 3 left over.

Part A: How many swings are there in all?

71 swings

Part B: Write a division problem to model this situation.

71 ÷ 17 = 4 R 3

19. Write *always*, *sometimes*, or *never* for each of the following statements.

The remainder is less than the divisor. ___ always

The quotient is greater than the remainder. ___ sometimes

The divisor is equal to the dividend. ___ sometimes

20. A charity has raised $14,569 for use in food banks around the country. There are 17 different food banks that will split the funds. How much money does each food bank receive?

$857

NAME _____ DATE _____ SCORE _____

Chapter 5 Test

1. Write a number that rounds to 2.3 when rounded to the nearest hundredth *and* when rounded to the nearest tenth, but does not round to 2.3 when rounded to the nearest thousandth.

Sample answer: 2.2992

2. Garret's teacher asked him to round the following number to the nearest tenth. Write Garret's answer in both expanded form and standard form.

$$3 \times 10 + 4 \times 1 + 2 \times \frac{1}{10} + 4 \times \frac{1}{100}$$

Expanded Form

$3 \times 10 + 4 \times 1 + 2 \times \frac{1}{10}$

Standard Form

34.2

3. Maria claims that it does not matter whether she first rounds two numbers to the nearest hundredth and then adds them or whether she first adds the two numbers and then rounds to the nearest hundredth. Is she correct? Explain.

Maria is not correct. Sample answer: If she rounds 4.566 and 1.228 to the nearest hundredth, then she gets 4.57 and 1.23. The sum is 5.80. However, if she first adds them, she gets 5.794. Rounded to the nearest hundredth, this is 5.79. The answers are different.

4. Veronica is buying the following items from the grocery store.

Milk	$2.60
Bread	$1.80
Crackers	$3.51
Cheese	$6.78

Part A: Round each term to the nearest dollar to estimate the total bill.

$16.00

Part B: Is the estimate greater than or less than the exact total? How do you know?

Greater than; every amount was rounded up to the nearest dollar.

5. Ophelia and her father took a three-day bike trip. On the first day, they rode 27.5 miles. On the second day they rode 23.2 miles. The total for the trip was 65.7 miles. Ophelia wants to know about how long the ride was on the third day. Is this a question about an exact answer or an estimate? Answer Ophelia's question.

Estimate. About 15 miles.

6. Joaquin used 10-by-10 grids to model 1.79 + 1.36. He is stuck. Describe what Joaquin must do next, then find the sum.

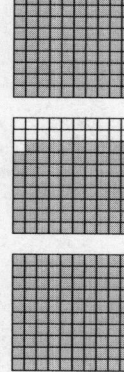

Joaquin needs to regroup twice. 1 hundredth from the 1.36 needs put with the 9 hundredths in the 1.79. Then 2 tenths from the 1.36 need put with the 1.79. The answer is 3.15.

Chapter Tests

Chapter 5 Test

7. Look at the following model. Circle the expressions that could match the model.

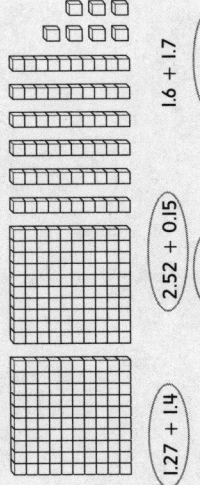

- (1.27 + 1.4) *(circled)*
- (2.52 + 0.15) *(circled)*
- 1.6 + 1.7
- 1.66 + 1.1
- (1 + 1.67) *(circled)*
- (1.07 + 1.60) *(circled)*

8. Tax on two purchases was $0.36 and $0.87.

Part A: Shade the model to show the regrouping needed to find the sum of the two values.

Part B: What is the sum?

1.23

9. On a class trip to Washington D.C. the bus made four stops for gas and spent the following amounts. What was the total gas bill for the trip?

$123.57
$135.88
$132.19
$98.27

$489.91

10. Kiara bought two items from a music store. The first item was between $8.00 and $9.00. The total of the items was $20.21. Write down two different possibilities for the cost of each item.

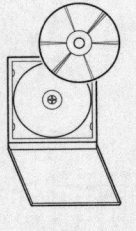

$8.78 and $11.43	$8.20 and $12.01

Sample answer.

11. Shade the box under "Yes" or "No" to indicate whether each problem will require regrouping.

Yes	No	
☐	▨	3.61 + 0.71
▨	☐	41.23 + 21.32
☐	▨	33.51 + 33.5
▨	☐	4.09 + 12.5

12. Write the correct property of addition for each step.

$9.9 + (3.6 + 4.1) + 0$

$= 9.9 + (4.1 + 3.6) + 0$ ___Commutative Property___

$= (9.9 + 4.1) + 3.6 + 0$ ___Associative Property___

$= 14 + 3.6 + 0$ ___Addition___

$= 17.6 + 0$ ___Addition___

$= 17.6$ ___Identity Property___

13. Circle the problems that will require regrouping.

- (13.71 − 2.8) *(circled)*
- (65.67 − 13.91) *(circled)*
- 3.4 − 2.2
- 245.16 − 5.16
- 145.65 − 140.05
- (123.45 − 11.111) *(circled)*

Chapter 5 Test

14. The local college baseball stadium recorded the number of people in attendance at the first three games of the season.

Game 1	14,998 people
Game 2	10,672 people
Game 3	15,002 people

Part A: Write down the best order in which to add the numbers so that it is easiest to find the total using mental math.

14,998 + 15,002 + 10,672

Part B: Find the total attendance for the first three games.

40,672 people

15. Write and solve a word problem for the following place-value chart.

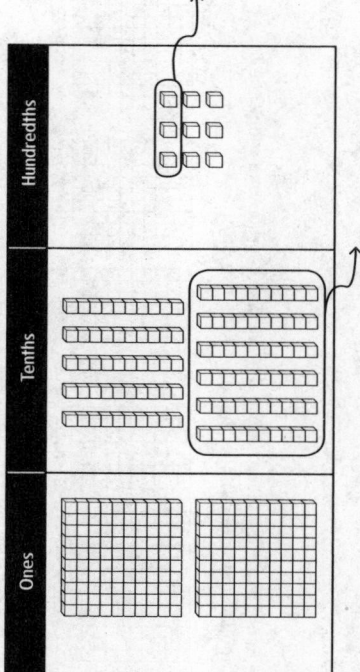

Ones	Tenths	Hundredths

Sample answer: Shane had $3.19. He spent $0.63. How much does Shane have left? The answer is $2.56.

16. A woodworker has purchased 92 linear meters of wood for framing. Each frame takes 18.21 linear meters of wood to make. Fill in the following chart to determine how many frames the woodworker can make with 92 linear meters of wood and how much wood will be left over.

Frames Made	Wood left over
1	92 − 18.21 = 73.79 meters
2	55.58 meters
3	37.37 meters
4	19.16 meters
5	0.95 meters

5 frames

0.95 meters left over

17. A baker needs to have 15.5 pounds of flour on hand for the weekend bread sale. He looks in the cupboard and finds a bag that has 4.6 pounds, a bag that has 6.1 pounds, and a bag that has 4.7 pounds. Does the baker have enough? If so, how much extra does he have? If not, how much more does he need?

He does not have enough. The baker still needs 0.1 pounds of flour.

18. Fill in the following missing digits so that the subtraction problem is accurate.

```
  3 1 . 4 2 6
- 2 0 . 7 1 7
-----------
  1 0 . 7 0 9
```

NAME _____ DATE _____

SCORE _____

Chapter 6 Test

1. The number of people who bought admission to the community pool last Friday was 198. Admission was $4.95. Estimate the amount of money that the pool brought in.

$1,000

2. Tamar walks 0.3 miles to school every day. Regroup and shade the models to figure out how far he walks in five days.

Monday   Tuesday   Wednesday   Thursday   Friday

1.5 _____ miles

3. Sort each problem into the box that lists the correct number of decimal places in the answer.

34.15 × 12   4.67 × 11   4.56 × 12,345

0.45 × 23   4 × 2.5   45 × 123.4

0 Decimal Places	1 Decimal Place	2 Decimal Places
4 × 2.5	34.15 × 12	4.67 × 11
45 × 123.4	4.56 × 12,345	0.45 × 23

4. Bihn is selling the following items at a garage sale.

CDs	$1.25
Bike	$25.16
Books	$0.76
Skates	$9.12

Part A: How much will he make if he sells 5 CDs, one bike, 15 books, and 2 pairs of skates?

$61.05

Part B: Bihn intends to purchase a $65 video game with his earnings. How much more money does he need if he sells all of the items in Part A?

$3.95

5. Mr. Olson is building a new doll house for his daughters. The base of the doll house will be 1.8 meters by 0.7 meters.

Part A: Shade the region of the base in the blocks below.

Part B: Rearrange the shading to help figure out the area of the base.

1.26 _____ square meters

6. A runner can run a mile in 6.27 minutes. If the runner maintains this pace, how many minutes will it take the runner to run 26.2 miles?

164.274 minutes

7. The amount of sugar in a serving of each of three brands of cookies is shown in the table. Which contains more grams of sugar: 1 serving of Brand A, 1.5 servings of Brand B, or 2 servings of Brand C?

Brand	Grams of Sugar per Serving
A	33.13
B	22.6
C	15.6

1.5 servings of Brand B

8. Malik is tiling his kitchen floor. The kitchen measures 13.25 feet by 11.75 feet.

Part A: What is the area that needs tiled?

155.6875 square feet

Part B: Tile comes in packs that will cover 10 square feet. How many packs will Malik need to buy?

16 packs

9. Put the following numbers in order from least to greatest.

1.23×10^3 123.0×10^2 0.0123×10^4

0.0123×10^4, 1.23×10^3, 123.0×10^2

10. Yasmine reads 12 pages of her book the first day, 18 pages the second day, 24 pages the third day, and so on. If the pattern continues, on what day will Jasmine finish her 264 page book?

On the eighth day

11. Harris is asked by his teacher to multiply the following in his head.

$(2.5 \times 7) \times (2 \times 4) \times 50$

Part A: Rewrite the expression so that the multiplication is easier to do in your head.

2.5 × 4 × (2 × 50) × 7

Part B: Find the answer.

7000

12. Light bulbs come in a pack of 6 that costs $12.29. Estimate the price per light bulb.

$2.00 per light bulb

13. Santiago was given the following diagram that is supposed to represent a decimal division. Complete the division equation.

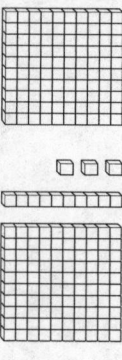

3.39 ÷ 3 = 1.13

Chapter 6 Test

14. Hot dogs are sold in packs of various sizes. Which of the three brands is the best buy?

Brand	Number in Pack	Price
A	8	$4.48
B	10	$5.56
C	12	$6.12

Brand C

15. A jeweler purchased 1.8 feet of gold chain to make bracelets. If each bracelet requires 0.6 feet, draw a model to help find how many bracelets he can make.

_3_ bracelets

16. The area of a picture frame is 16.625 square feet. The length is 3.5 feet. Find the width.

4.75 feet

17. Daniel claims that a decimal divided by a decimal can never be a whole number. Is he correct? If so, explain why. If not, give an example showing that he is wrong.

Daniel is not correct. 2.6 ÷ 1.3 = 2, which is a whole number.

18. On Monday, a driver purchased 18 gallons of gas for $61.38.

Part A: What was the price of gas per gallon that day?

$3.41 per gallon

Part B: The next day, the price of gas was $3.29 per gallon. How much could the driver have saved had he waited until Tuesday to buy gas?

$2.16

19. Circle the expression that is not equal to the others.

567 ÷ 1,000   5.67 ÷ 10

56.7 ÷ 10,000   56.7 ÷ 100

20. In July of 2014, Phoenix, Arizona had a record rainfall of 5.04 inches in 12 hours. On average, how many inches of rain fell per hour?

0.42 inches per hour.

The expression circled is 56.7 ÷ 10,000.

Grade 5 · Chapter 6 Multiply and Divide Decimals   87

88   Grade 5 · Chapter 6 Multiply and Divide Decimals

230   *Grade 5* · **Chapter 6** Multiply and Divide Decimals

Copyright © McGraw-Hill Education. Permission is granted to reproduce for classroom use.

NAME _____ DATE _____ PERIOD _____

SCORE _____

Chapter 7 Test

1. Gary watches cars go by his house. He counted 4 red cars, 3 blue cars, 4 yellow cars, 4 white cars, 3 black cars, and 4 green cars.

Part A: Write and evaluate an expression for the total number of cars Gary saw using only addition.

$$4 + 3 + 4 + 4 + 3 + 4 = 22 \text{ cars}$$

Part B: Write and evaluate an expression for the total number of cars Gary saw using both addition and multiplication.

$$(4 \times 4) + (2 \times 3) = 22 \text{ cars}$$

2. A football is thrown up in the air. The height of the football after two seconds is $3 \times 2^2 + 4 \times 2 + 6$. Find the height of the football.

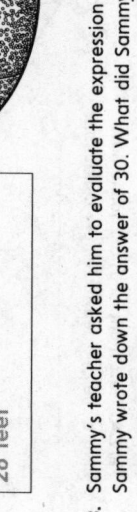

26 feet

3. Sammy's teacher asked him to evaluate the expression $2 + 4 \times 5$. Sammy wrote down the answer of 30. What did Sammy do wrong? Give the correct answer.

Sammy added the 2 and the 4 first. The order of operations says to do the multiplication first. The correct answer is 22.

---

4. Camilla and her three friends bought three tickets to the movie theater at $8 each. They also bought a large popcorn for $10. They split the bill evenly. Write and evaluate an expression for the total amount that each friend spent.

$$[(4 \times 8) + 10] \div 4; \ \$10.50 \text{ each}$$

5. Jared saved $125 over the course of the last month. His friend Hector saved twice the difference between Jared's amount and $50. Circle which of the following expressions represents the amount that Hector saved in the last month.

$2 \times 125 - 50$

$(50 - 2) \times 125$

$\boxed{2 \times (125 - 50)}$

$2 \times 50 - 125$

6. The cost of admission to the county fair is $12. The cost of each ride at the fair is $2. John went to the fair and rode 14 rides.

Part A: Write an expression for how much John spent.

$12 + 2 \times 14$

Part B: Evaluate the expression to find out how much John spent.

$40

Chapter Tests

7. A pizza shop sells a single pizza for $11. However, at the end of the night, they will sell the extra pizzas for $6 each. If the pizza shop sold 51 pizzas before the discounted price and made $615, find the number of discounted pizzas they sold.

9 pizzas

8. Joline decided to sell her china doll collection. She sold the dolls for $24 each. Joline sold 8 dolls to Sylvia, and then sold half of the remaining dolls to Jane. She made $288.

Part A: How many dolls did Joline have before she decided to sell her collection?

16 dolls

Part B: How many dolls does Joline have now?

4 dolls

9. In which pattern will the numbers go over 100 first? Write out the patterns to show your answer.

Pattern A: Start at 2 and multiply by 2.

Pattern B: Start at 55 and add 10.

Pattern A: 2, 4, 8, 16, 32, 64, 128
Pattern B: 55, 65, 75, 85, 95, 105
Pattern B goes over 100 first.

10. Look at the number of blocks in the following pattern.

Part A: Write down the pattern for the number of blocks in each stack.

Start with 1. Then add 2. Then add 3. Then add 4. And so on.

Part B: Predict how many blocks will be in the next stack.

15 blocks

11. Circle the pattern that does not belong.

3, 9, 27, 81

5, 15, 45, 135

2, 6, 18, 54

2, 5, 8, 11 (circled)

12. Mrs. Gerard saves $10 per week. Mr. Gerard saves $8 per week. Fill in the following table for the total amount in savings at the end of each week, and describe the pattern for the total savings.

Week	Mrs. Gerard	Mr. Gerard	Total Savings
1	$10	$8	$18
2	$20	$16	$36
3	$30	$24	$54
4	$40	$32	$72

Pattern: Add $18

13. A park ranger is taking a tour of the major sites in a national forest to check for safety violations. The map below shows the sites.

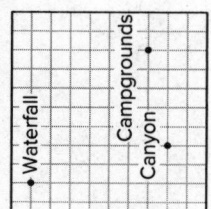

Part A: If the ranger starts at the canyon, describe the path he can take first to the waterfall, and then to the campgrounds if he must follow the grid lines.

2 units west, then 7 units north to the waterfall. Then 7 units east, followed by 6 units south to the campgrounds.

Part B: If the ranger wants to go directly from the canyon to the campgrounds, how many units shorter is that than going to the waterfall first? Again, the ranger must follow the grid lines.

16 units shorter

14. Brock is trying to make a plan for building a table. He wants the top to be a rectangle. He has placed three of the corners on the coordinate plane below.

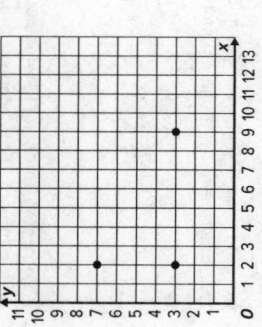

What is the coordinate of the fourth corner? _____ (9, 7)

---

15. Jackie and Taylor decide to start exercise routines. Jackie does 5 situps on the first day and adds 2 each day. Taylor does 3 situps on the first day and adds three each day.

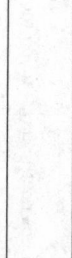

Part A: Find the number of situps that each girl does for the first four days, and graph them as ordered pairs.

Part B: On which day does each girl perform the same number of situps?

Day 3

16. Mr. Zhao walks from the grocery store to his house in a straight line. The grocery store is at point (1,2). His house is at point (11,7). For the ordered pairs listed, circle all of the order pairs that Mr. Zhao passes through on his way home.

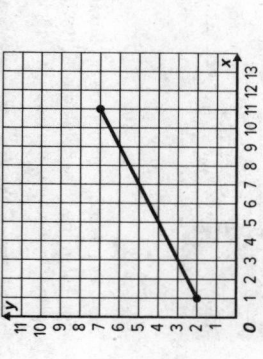

(3, 3) (4, 5) (7, 5) (9, 6) (10, 7)

NAME

DATE

SCORE

Chapter 8 Test

1. Mr. King used 16 gallons of gas in 5 days. Circle any of the following that describe the average amount of gas Mr. King used per day.

   $\dfrac{16}{5}$ $\dfrac{5}{16}$ $3\dfrac{1}{5}$ $5\dfrac{1}{3}$

2. A carpenter is framing 12 windows in a house. He used 51 feet of wood on the project. If the windows are all the same size, how many feet of wood were used in a single window? Write the answer in two different ways using fractions.

   $\dfrac{17}{4}$ feet or $4\dfrac{1}{4}$ feet

3. Consider the following five numbers.

   12   ~~15~~   36   18   ~~8~~

   Part A: What is the greatest common factor of all five numbers?

   1

   Part B: Cross out two of the five numbers so that the remaining numbers have a greatest common factor of 6.

4. The table below shows the number of flowers that a florist has to put in vases for a wedding display. Each vase will have only one color of flower, and the florist wants to make sure that each vase has the same number of flowers. If the florist puts all of the flowers in vases, what is the greatest number of flowers that could be in each vase?

Pink	18
Red	30
White	24

   6 flowers

5. Deandre has made a table of his baseball card collection based on the year. Fill in the third column with the fraction of his total cards that year represents, and put the fractions in lowest terms.

Year	Number of Cards	Fraction of Total Cards
2009	12	$\dfrac{1}{5}$
2010	18	$\dfrac{3}{10}$
2011	10	$\dfrac{1}{6}$
2012	20	$\dfrac{1}{3}$

6. Shade the box next to any fraction that is in simplest form.

   ☐ $\dfrac{8}{12}$ ☐ $\dfrac{9}{12}$

   ☐ $\dfrac{10}{12}$ ■ $\dfrac{11}{12}$

7. Dai went shopping and received $1.55 in change in quarters and dimes. He told his friend Ginny that the change came in 11 coins and asked her to guess how many of each coin he had. Ginny guessed that there were 5 quarters and 3 dimes.

Part A: Is Ginny correct? Explain

> Ginny is not correct. Her guess adds up to $1.55, but her guess included only 8 coins, and Dai had 11.

Part B: If Ginny's guess is not correct, find the correct answer.

> 3 quarters and 8 dimes

8. Mrs. Franklin took her five children to an amusement park. The cost of tickets for the 2 younger children was $12.50 each. The cost of her ticket was $15.00. She spent a total of $80.50. What was the cost of each ticket for her three older children?

> $13.50

9. Isabelle buys gas every five days. If she buys gas today and today is a Saturday, how many more days will it be before she buys gas on a Saturday again?

> 35 days

---

10. Mr. Sanchez goes to the movies every 18 days. His brother goes to the movies every 30 days. If they were at the movies together this evening, how many more days will it be before they are at the movies together again?

> 90 days

11. Students from three different fifth grade math classes were asked to return permission slips for a field trip. The table shows what fraction of each class turned in their permissions slips on the first day. Number the classes in order from 1 to 3 starting with the class that had the smallest fraction of students turning in the slips.

Mrs. Haley's class	$\frac{5}{12}$
Mr. Black's class	$\frac{5}{11}$
Mrs. Mayne's class	$\frac{2}{5}$

2 Mrs. Haley's class

3 Mr. Black's class

1 Mrs. Mayne's class

12. The local football coach has always had a goal of winning $\frac{3}{4}$ of the games in a season. The numbers below show the fraction of games won for the season. Circle those seasons below when the coach met his goal.

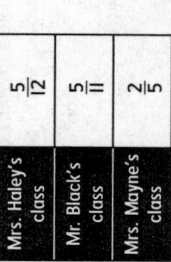

2011-2012 Season $\frac{8}{11}$

2012-2013 Season $\frac{10}{13}$ (circled)

2013-2014 Season $\frac{10}{14}$

Chapter Tests

17. The table shows the amount of rainfall that a town experienced in the last three months.

April	10 inches
May	9 inches
June	6 inches

Part A: What fraction of the total rainfall came in April? Put your answer in lowest terms.

$\frac{2}{5}$

Part B: What fraction of the total rainfall came in May? Write your answer as a decimal.

0.36

18. Jaylon is asked to find the greatest common factor of three numbers. He finds the greatest common factor of the first two numbers to be 1. Jaylon claims he knows the answer without any extra work. How does he know?

1 is a factor of every number. If the greatest common factor of the first two numbers is 1, then there is no other factor that could divide into all three numbers. So the greatest common factor of all three numbers is 1.

13. A baseball player got 7 hits in his last 25 at bats. Shade the model to help write the decimal that represents the fraction of at bats resulting in a hit.

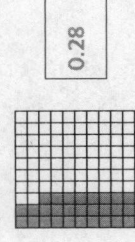

0.28

14. Circle all of the numbers that could be modeled with the following block.

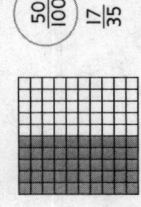

$\frac{10}{20}$ 0.5

$\frac{50}{100}$ $\frac{17}{35}$

15. Match each fraction with its decimal equivalent.

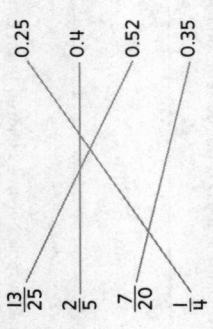

0.25
0.4
0.52
0.35

$\frac{13}{25}$
$\frac{2}{5}$
$\frac{7}{20}$
$\frac{1}{4}$

16. Felipe built a model train that has a scale of $\frac{21}{25}$. Write this number as a decimal.

0.84

NAME _____ DATE _____ PERIOD _____

SCORE _____

Chapter 9 Test

1. Frederick has been asked to sort five bags into three categories based on whether they are closest to "empty", "half full", or "full". Each fraction represents how full the bag is. Write each of the fractions in the correct box.

$\frac{2}{19}$ $\frac{2}{5}$ $\frac{2}{7}$ $\frac{6}{11}$ $\frac{10}{12}$

About Empty	About Half Full	About Full
$\frac{2}{19}$	$\frac{2}{7}$ $\frac{6}{11}$	$\frac{10}{12}$

2. The following chart shows how much it snowed in the first three hours of a snowstorm. Find the total snowfall in inches for the first three hours. Write your answer in lowest terms.

First Hour	$\frac{1}{8}$ inch
Second Hour	$\frac{3}{8}$ inch
Third Hour	$\frac{2}{8}$ inch

$\frac{3}{4}$ inch

3. Tamika bought red beans, pinto beans, and black beans. She purchased $\frac{15}{16}$ pounds of beans in total. If Tamika bought $\frac{3}{16}$ pounds of red beans and $\frac{5}{16}$ pounds of pinto beans, find the weight of black beans she purchased. Write your answer in lowest terms.

$\frac{7}{16}$ pound

---

4. Keisha bought $\frac{1}{6}$ tank of gas in the morning on her way to work. She added $\frac{3}{4}$ tank on her way home from work. Fill in the model below to determine what fraction of a tank she bought altogether.

$\frac{1}{4}$	$\frac{1}{4}$	$\frac{1}{4}$	$\frac{1}{4}$	$\frac{1}{6}$
$\frac{1}{12}$ $\frac{1}{12}$ $\frac{1}{12}$	$\frac{1}{12}$ $\frac{1}{12}$ $\frac{1}{12}$	$\frac{1}{12}$ $\frac{1}{12}$ $\frac{1}{12}$	$\frac{1}{12}$ $\frac{1}{12}$	$\frac{1}{12}$

$\frac{11}{12}$ tank

5. Feng was asked by his teacher to add $\frac{1}{5}$ and $\frac{3}{10}$. Feng got an answer of $\frac{6}{5}$.

Part A: Compare the two fractions to $\frac{1}{2}$ and show that Feng cannot be correct.

$\frac{1}{5}$ and $\frac{3}{10}$ are both less than $\frac{1}{2}$, so their sum must be less than 1. $\frac{6}{5}$ is greater than 1. Feng cannot be correct.

Part B: What is the correct answer in lowest terms?

$\frac{1}{2}$

6. Circle which of the following does not belong, and describe why.

$\frac{1}{3} + \frac{1}{2}$ $\left(\frac{1}{6} + \frac{1}{6}\right) + \frac{1}{2}$ $\frac{1}{3} + \left(\frac{1}{6} + \frac{1}{6} + \frac{1}{6}\right)$ $\boxed{\left(\frac{1}{2} + \frac{1}{2} + \frac{1}{2}\right) + \frac{1}{2}}$

All of the other sums equal $\frac{5}{6}$.

---

7. Mrs. Prim had $\frac{3}{5}$ of a cup of flour. She used $\frac{1}{2}$ cup on a dessert recipe.

Part A: Fill in the model below to help find how much flour Mrs. Prim has left.

$\frac{1}{5}$	$\frac{1}{5}$	$\frac{1}{5}$
$\frac{1}{2}$	$\frac{1}{5}$	$\frac{1}{10}$

$\frac{1}{10}$ ___ cup

Part B: Mrs. Prim needs another $\frac{1}{4}$ cup for a sauce recipe for the dessert. Does she have enough flour left? Explain.

No. $\frac{1}{10} < \frac{1}{4}$.

8. Luciano walked $\frac{7}{18}$ mile on Saturday morning. On Sunday, she walked $\frac{5}{9}$ mile. How much more did she walk on Sunday than on Saturday? Shade the box next to any correct answer.

☐ $\frac{3}{18}$ ☐ $\frac{1}{6}$ ☐ $\frac{2}{12}$

9. Fill in the box to make a true number sentence.

$$\frac{11}{12} - \frac{\boxed{2}}{3} = \frac{1}{4}$$

10. A field goal kicker practices by moving back the same number of yards every time he kicks. On the third kick he is 40 yards away from the goal posts. On the sixth kick he is 49 yards away. Fill in the following chart and work backwards to figure out how far away the kicker was on the first kick. Circle the answer.

Kick #1	(34 yards)
Kick #2	37 yards
Kick #3	40 yards
Kick #4	43 yards
Kick #5	46 yards
Kick #6	49 yards

11. Mia has three colors of fabric: red, blue, and purple. She has $3\frac{7}{8}$ yards of red fabric and $10\frac{2}{15}$ yards of blue fabric. If she has $19\frac{11}{13}$ yards of fabric al together, about how much purple fabric does she have? Show how she estimated.

6 yards; $20 - (4 + 10)$

12. Write and solve a problem involving addition and mixed numbers. Put all mixed numbers in lowest terms.

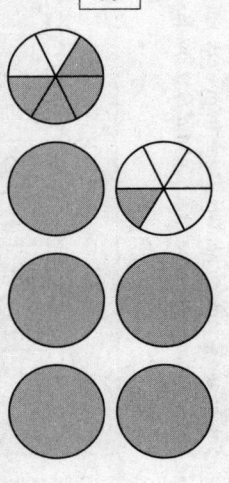

$$3\frac{2}{3} + 2\frac{1}{6} = 5\frac{5}{6}$$

13. Jasmine is making a fruit smoothie that requires the following amounts of certain juices.

Apple Juice	$3\frac{1}{4}$ cups
Raspberry Juice	$\frac{3}{8}$ cup
Grape Juice	$\frac{1}{2}$ cup

Part A: Find the total amount of juice that this recipe will make.

[$4\frac{1}{8}$ cup]

Part B: If Jasmine already has $1\frac{1}{2}$ cups of apple juice, how much more will she need to buy in order to have enough for the recipe?

[$1\frac{3}{4}$ cup]

14. A construction crew is painting lines on the side of a new highway. In one week they are supposed to have $67\frac{1}{2}$ miles completed. On Monday, they painted $13\frac{5}{6}$ miles, and on Tuesday they painted $12\frac{1}{3}$ miles. How many more miles are left to paint?

[$41\frac{1}{3}$ miles]

15. Circle which problems would require renaming.

 $4\frac{1}{3} - 2\frac{1}{2}$ $5\frac{5}{6} - 3\frac{1}{3}$ 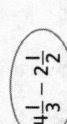 $12\frac{3}{8} - 7\frac{2}{5}$

16. Julian has three lengths of rope: $12\frac{1}{8}$ feet, $13\frac{3}{8}$ feet, and $11\frac{7}{8}$ feet.

Part A: Estimate the total length of rope that Julian has by rounding to the nearest whole foot.

[37 feet]

Part B: Estimate the total length of rope by rounding to the nearest half foot.

[$37\frac{1}{2}$ feet]

Part C: Find the exact length of rope that Julian has in total.

[$37\frac{3}{8}$ feet]

17. In the following model fill in the tiles with fractions that make the quantities equal.

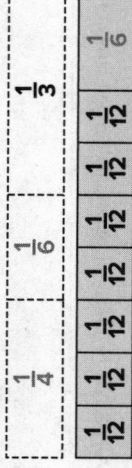

18. Rashaun purchased a 5 gallon bucket of paint. He spilled some paint when opening the bucket. There are $3\frac{3}{8}$ gallons left in the bucket. How much paint did Rashaun spill?

[$1\frac{5}{8}$ gallons]

Chapter Tests

NAME _____ DATE _____

SCORE _____

Chapter 10 Test

1. Santana took out 24 books from the library. She returned $\frac{3}{8}$ of them.

Part A: Draw a model to illustrate the situation.

Part B: How many books does Santana have left?

15 books

2. Yolanda says that $\frac{3}{5}$ of 2 is the same as $\frac{2}{5}$ of 3. Draw bar models to show that she is correct.

3. A cook has $11\frac{7}{8}$ cups of flour. He uses $\frac{2}{3}$ of his flour on Saturday morning. Estimate how many cups he used. Show how you estimated.

$\frac{2}{3} \times 12 = 8$ cups

4. Look at the following model.

Part A: Shade the blocks on the right side of the equal sign so that the model represents a true statement.

Part B: Write a multiplication problem and answer for the model.

$\frac{2}{3} \times 3 = 2$

5. A carpenter bought 11 linear feet of an oak board. He used $\frac{2}{5}$ of the board on a baseboard. How many feet did he use?

$4\frac{2}{5}$ feet

6. Thelma has 8 yards of fabric. She used $\frac{2}{3}$ of the fabric on a project. Circle any answer that shows how many yards Thelma has left.

$\frac{16}{3}$ yards $5\frac{1}{3}$ yards

$\frac{8}{3}$ yards $2\frac{2}{3}$ yards

Chapter 10 Test

7. A field is $\frac{2}{3}$ miles long and $\frac{1}{2}$ mile wide. Draw a model to help find the area of the field.

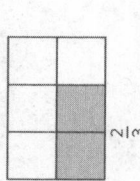

Area = $\frac{1}{3}$ square mile

8. An electrician has $\frac{14}{15}$ yard of wire. He used $\frac{5}{7}$ of the wire.

Part A: How many yards of wire did the electrician use?

$\frac{2}{3}$ yard

Part B: How many yards of wire does the electrician have left?

$\frac{4}{15}$ yard

9. The area of a triangle can be found by multiplying $\frac{1}{2}$ times the base times the height. Find the area of the triangle.

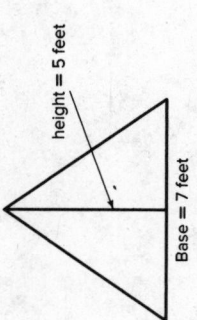

height = 5 feet

Base = 7 feet

Area = $17\frac{1}{2}$ square feet

10. Jentilly bought $3\frac{2}{3}$ pounds of black cherries. She bought $1\frac{1}{2}$ times that amount in red cherries.

Part A: How many pounds of red cherries did Jentilly buy?

$5\frac{1}{2}$ pounds

Part B: How many pounds of cherries did Jentilly buy in total?

$9\frac{1}{6}$ pounds

11. Fill in the circles with <, >, or =.

$10 \times \frac{2}{3}$ ⟨<⟩ 10

$17 \times \frac{7}{4}$ ⟨>⟩ 17

$12 \times 1\frac{2}{5}$ ⟨>⟩ 12

$8 \times \frac{9}{9}$ ⟨=⟩ 8

12. Kaitlin spent $\frac{2}{3}$ of an hour on homework. Her sister, Judy, spent $1\frac{1}{2}$ times that amount. Explain why Judy's homework time is between $\frac{2}{3}$ of an hour and $1\frac{1}{2}$ hour.

Judy's homework time is $\frac{2}{3} \times 1\frac{1}{2}$. Since $\frac{2}{3} < 1$, this product is less than $1\frac{1}{2}$. Since $1\frac{1}{2} > 1$, the product is greater than $\frac{2}{3}$.

Chapter Tests

16. Gretchen claims that $\frac{1}{3} \div 4$ is the same as 4 divided by $\frac{1}{3}$. Is she correct? Explain.

No, she is not correct. $\frac{1}{3} \div 4 = \frac{1}{12}$, but $4 \div \frac{1}{3} = 12$.

17. Ishmael bought a brand new bag of green, yellow, and red marbles. There are 40 marbles in the bag. $\frac{1}{4}$ of the marbles are yellow. The number of green marbles is the same as the number of red marbles. How many of each color are there?

15 red marbles

10 yellow marbles

15 green marbles

18. Sort each of the fraction problems into those for which the product or quotient is greater than 1, those for which the product or quotient is less than 1, and those for which the product or quotient is equal to 1.

$\frac{1}{4} \div 3$ $2 \div \frac{1}{5}$ $2\frac{1}{5} \times 3$

$\frac{1}{7} \times 7$ $3\frac{1}{6} \times \frac{1}{2}$ $\frac{1}{8} \times 2$

Greater than 1	Less than 1	Equal to 1
$2 \div \frac{1}{5}$	$\frac{1}{4} \div 3$	$\frac{1}{7} \times 7$
$2\frac{1}{5} \times 3$	$\frac{1}{8} \times 2$	
$3\frac{1}{6} \times \frac{1}{2}$		

13. Chase needs to cut his fishing line into pieces that are $\frac{1}{4}$ of a foot. He has 3 feet of fishing line. Use fraction tiles to help figure out how many pieces Chase can make.

$\frac{1}{4}$	$\frac{1}{4}$	$\frac{1}{4}$	$\frac{1}{4}$
$\frac{1}{4}$	$\frac{1}{4}$	$\frac{1}{4}$	$\frac{1}{4}$
$\frac{1}{4}$	$\frac{1}{4}$	$\frac{1}{4}$	$\frac{1}{4}$

12 pieces

14. Mrs. Chen slices 5 pies. Each slice represents $\frac{1}{8}$ of a pie. How many slices of pie are there?

40 slices

15. Tyrone has $\frac{1}{6}$ pound of almonds. He wants to split this equally among 5 people. What fraction of a pound will each person get?

$\frac{1}{30}$ pound

NAME _____ DATE _____ SCORE _____

Chapter 11 Test

1. Chelsea is measuring a piece of thread.

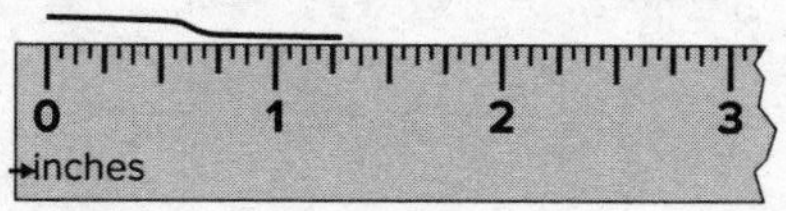

Part A: What is the length measured to the nearest half inch?

$1\frac{1}{2}$ inches

Part B: What is the length measured to the nearest quarter inch?

$1\frac{1}{4}$ inches

2. A biologist determined the wingspan of a baby bird to be $6\frac{3}{4}$ inches when measured to the nearest quarter inch. Her lab partner measured the same wingspan to the nearest half inch and got 7 inches. Circle the statement that describes the actual wingspan.

Less than $6\frac{3}{4}$ inches   ⟨Between $6\frac{3}{4}$ and 7 inches⟩

Greater than 7 inches

3. Compare using <, >, or =.

144 in. = 12 ft   6.5 yd > 18 ft

2 mi < 10,561 ft   72 in. = 2 yd

---

4. A runner ran a marathon, which is 26.2 miles. How many feet is this?

138,336 feet

5. Mr. Schnur is carpeting his living room. The dimensions are show in the diagram. However, the carpet company wants to know how many square inches this is. Find the area of the floor in square inches.

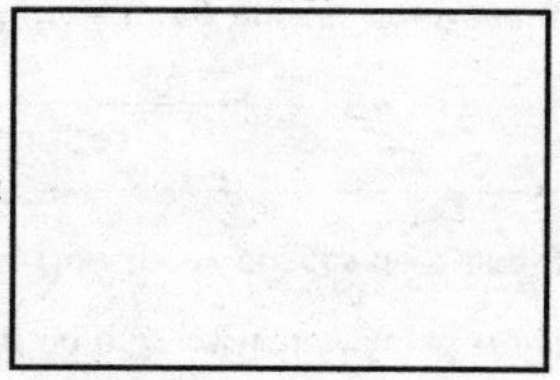

17 feet

11 feet

26,928 square inches

6. Francis measured the weight of a red block and got 5 ounces. He then found that a green block weighs the same as 4 red blocks.

Part A: What is the weight of the green block in ounces?

20 ounces

Part B: What is the weight of the green block in pounds and ounces?

1 pound 4 ounces

Part C: What is the weight of the green block in pounds?

$1\frac{1}{4}$ pounds

Chapter 11 Test

7. How much heavier is 7 pounds than 110 ounces?

2 ounces

8. Frederico has three bags of flour with the following weights.

Bag 1	15 ounces
Bag 2	14 ounces
Bag 3	11 ounces

How many pounds of flour does Frederico have in all?

2.5 pounds

9. A car manufacturer makes two models. Model A weighs 1.5 T. Model B weighs 2,900 pounds.

Part A: Which model weighs more?

Model A

Part B: What is the difference between the two weights in pounds?

100 pounds

10. Circle the statements that are false.

7 pints > 14 cups | 2 gallons > 17 pints
9 cups > 2 quarts | 1 gallon = 16 cups

11. Calisto is making tea for a tea party. Each serving will be 1 cup. How many pints of tea will she need for 16 guests?

8 pints

12. Dr. Blanchard recommends drinking a gallon of water a day. His patient, Sam, has an 8 fluid ounce glass. How many glasses does Sam need to drink a day in order to follow the doctor's recommendation?

16 glasses

13. One gallon of a particular liquid weighs $8\frac{1}{4}$ pounds.

Part A: How many ounces does one gallon of the liquid weigh?

132 ounces

Part B: How many ounces does one quart weigh?

33 ounces

Part C: How many ounces does one pint weigh?

16.5 ounces

Chapter 11 Test

16. A small table for a dollhouse requires 6 cm of a thin wooden board. A dollhouse maker has 12 meters of wood. How many tables can he make?

200 tables

17. A large rectangular field has length 1.5 km and width 0.75 km. What is the area of the field in square meters?

1,125,000 square meters

18. Shanice weighed her collection of books as 12,014 grams. Her brother, Marquis weighed his books and got 13 kg. Who has the greater weight in books? By how much?

Marquis, by 986 grams

19. If a farmer has 386 grams of tomatoes, 671 grams of potatoes, 711 grams of zucchini, and 997 grams of yellow squash, find the total weight in vegetables in kilograms.

2.765 kg

20. Circle the quantity that is not equal to the others.

7.1 L   7,100 mL   7 L 100 mL   (1 L 700 mL)

14. The line plot shows the snowfall in inches for the last fourteen days in the town of Snowshoe.

Part A: Make a new line plot that shows the snow fall in feet.

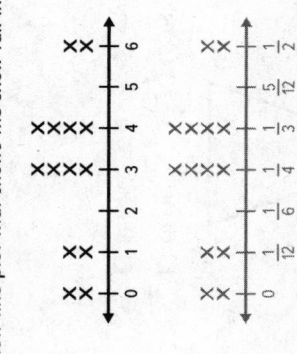

Part B: What is the fair share in feet if the same amount of snow fell every day?

$\frac{1}{4}$ feet

15. Match the item with the appropriate unit of measurement.

The length of a race — millimeter

The width of a postage stamp — centimeter

The thickness of a penny — meter

The length of a table — kilometer

Chapter Tests

NAME

DATE

Chapter 12 Test

SCORE

1. What shapes make up the surface of a soccer ball?

Regular pentagons and regular hexagons

2. A regular triangle has one side length of 15 cm. Find the perimeter of the triangle.

45 cm

3. A farmer is making a triangular corn maze for a fall attraction. Fill in the boxes next to the words that describe the triangle.

☑ acute ☐ obtuse ☐ right
☐ scalene ☑ isosceles ☐ equilateral

4. A construction company has marked off a site for a building in the shape of a quadrilateral. All four sides of the site are congruent. Opposite angles are congruent, but there are no right angles. Draw a possible shape for this site.

5. For each of the following pairs of quadrilaterals, describe one thing they have in common.

A rhombus and a square — Four equal sides

A trapezoid and a rectangle — At least one set of parallel sides

A square and a rectangle — Four right angles

A parallelogram and a rhombus — Two sets of parallel sides

6. Jorge says he is thinking of a quadrilateral that has two right angles and at least one set of parallel sides and says that it must be a rectangle.

Part A: Draw a picture that proves Jorge wrong.

Part B: Name your shape.

trapezoid

7. A box company is experimenting with nets for making their boxes. Circle any of the following that will fold up to make a box.

8. A bricklayer is laying bricks for a new patio.

Part A: Fill in the information below for the brick.

___6___ faces   ___12___ edges   ___8___ vertices

Part B: What is the name of the shape?

rectangular prism

9. Each cube has a side length of 1 inch. Find the volume of the prism.

24 cubic inches

10. A company that manufactures baskets makes them in the shape of a rectangular prism.

Part A: The volume of one basket is 29,750 cubic inches. The basket is 35 inches wide and 34 inches long. Find the height of the basket.

25 inches

Part B: The company makes two smaller baskets that have a volume of 9,600 cubic inches with a height of 25 inches. Find two possibilities for the length and the width.

Sample answer:
Length = 32 in, width = 12 in
Length = 8 in, width = 48 in

11. Number each of the shapes from 1 to 3 in increasing order according to their volume. Each cube is one cubic centimeter.

 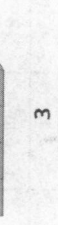 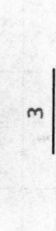 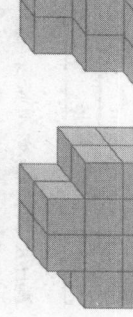

___1___   ___3___   ___2___

Chapter Tests

12. Mrs. Huan has 6 books in a stack. The bottom two books are 12 inches long, 1 inch thick, and 6 inches wide. The middle two books are 10 inches long, 1 inch thick, and 5 inches wide. The top two books are 7 inches long, 0.5 inches thick, and 4 inches wide. Find the volume of the stack.

272 cubic inches

13. A child is stacking blocks in a pyramid design as shown below, but much bigger. There are 66 blocks in all.

Part A: How many blocks are in the bottom row?

11 blocks

Part B: If each block is a cube with a side length of 2 inches, what is the volume of the construction?

528 cubic inches

14. Jackson's teacher asked him to look at what happens to the volume of a rectangular prism when every side is doubled.

Part A: Fill in the following chart.

Length	Width	Height	Volume
1 in	1 in	1 in	1 in³
2 in	2 in	2 in	8 in³
1 in	2 in	3 in	6 in³
2 in	4 in	6 in	48 in³
3 in	5 in	2 in	30 in³
6 in	10 in	4 in	240 in³

Part B: Describe what happens to the volume when each side is multiplied by 2.

The volume is multiplied by 8.

15. Mark each statement as true or false.

True	False	
☐	▨	All rectangles are squares.
▨	☐	Some parallelograms are rhombuses.
☐	▨	All rhombuses are regular quadrilaterals.
▨	☐	Some trapezoids are parallelograms.

16. An architect is building a house that has 5 faces, 9 edges, and 6 vertices. What is the name for a three-dimensional figure that has these characteristics?

triangular prism

Page 125 • Setting Goals

Task Scenario
Students will order whole numbers to create a yearly production goal for a factory that produces electronic components.

Depth of Knowledge	DOK2, DOK3	

Part	Maximum Points	Scoring Rubric
A	2	**Full Credit:** Sample answer: 1,423,411; 1,432,426; 1,432,501; 1,531,199; 1,532,199; 1,570,672; 1,600,121 Partial Credit (1 point) will be given for listing the numbers from greatest to least. No credit will be given for any other listing.
B	2	**Full Credit:** Sample answer: This goal is lower than any other year. Partial Credit (1 point) will be given for an answer that indicates that 1,423,000 is "too low", but fails to indicate that it would be the lowest in the last seven years. No credit will be given for an answer that does not indicate the low nature of the goal.
C	2	**Full Credit:** Sample answer: 1,532,200 units Partial Credit (1 point) will be given for an answer that is missing the label "units." No credit will be given for an incorrect answer.

Performance Task Rubrics

Part	Maximum Points	Scoring Rubric
D	2	**Full Credit:** Full credit may be given for an answer of "yes" or "no". Sample answer (no): No, my answer does not meet this requirement. It is less than 1,570,000 units. A better goal would be 1,570,500 units. This number is larger than 1,570,000 and still the third highest number. Sample answer (yes): Yes, my goal is larger than 1,570,000 units and is the third highest. Partial Credit (1 point) is offered in the "no" case where a student identifies that her goal does not meet the requirement, but does not offer another solution. It is offered in the "yes" case for a correct answer of "yes" that is missing an explanation. No credit is given for an incorrect answer.
E	2	**Full Credit:** Sample answer: The goal will meet this last requirement. My goal was 1,570,000 units, so the total for years 6, 7, and 8 is 4,525,110 units, which is greater than 4,500,000 units. Partial Credit (1 point) will be given for students who correctly identify that the goal will meet the third requirement but do not offer a reasonable explanation. No credit will be given for an incorrect answer.
TOTAL	10	

Performance Task (continued)

Part C

While the manager is tempted to set a new record by producing more units than have ever been produced in a year, he knows that people are not buying as many components as they used to, and he does not want to make more units than can be sold. He decides to set the goal of producing the third highest number of components in company history. Suggest a goal for the factory manager.

1,532,200 units

Part D

In researching the company financial reports, the factory manager discovers that the factory must produce at least 1,570,000 units in a year in order to make a profit. Does your goal from **Part C** meet this requirement? If so, explain why. If not, offer the factory manager a new goal that meets *both* requirements.

No, my answer does not meet this requirement. A better goal would be 1,570,500 units, which is larger than 1,570,000 and still the third highest.

Part E

The factory manager's supervisor indicates that it is absolutely essential that the total number of units sold in years 6, 7, and the new year 8 be at least 4,500,000. Explain why the goal you gave the factory manager in **Part D** will also meet this new requirement.

The goal will meet this last requirement. My goal was 1,570,500 units, so the total is 4,525,110 units, which is greater than 4,500,000 units.

Student Model

NAME _____   DATE _____

SCORE _____

Performance Task

Setting Goals

A factory produces electronic components. The new manager wants to set a goal for how many units will be produced in the upcoming year.

Write your answers on another piece of paper. Show all your work to receive full credit.

Part A

The factory has existed for seven years. The chart below gives the number of components produced by the factory each year.

Order

Year	Components		Order
1	1,432,426		2
2	1,532,199		5
3	1,432,501		3
4	1,570,672		6
5	1,600,121		7
6	1,423,411		1
7	1,531,199		4

The factory manager needs to put the data in order so that he can make a decision on the next year's goal. Order the data from least to greatest.

Part B

The factory manger asks his assistant manager to give input for the production goal. The assistant manager suggests 1,423,000 units. Explain why this goal may not be appropriate.

This goal is lower than any other year.

NAME _____ DATE _____

SCORE _____

Performance Task

Setting Goals

A factory produces electronic components. The new manager wants to set a goal for how many units will be produced in the upcoming year.

Write your answers on another piece of paper. Show all your work to receive full credit.

Part A

The factory has existed for seven years. The chart below gives the number of components produced by the factory each year.

Year	Components
1	1,432,426
2	1,532,199
3	1,432,501
4	1,570,672
5	1,600,121
6	1,423,411
7	1,531,199

The factory manager needs to put the data in order so that he can make a decision on the next year's goal. Order the data from least to greatest.

1,600,121
1,570,672
1,532,199
1,531,199
1,432,501
1,432,426
1,423,411

Part B

The factory manger asks his assistant manager to give input for the production goal. The assistant manager suggests 1,423,000 units. Explain why this goal may not be appropriate.

The goal is lower than some other years.

Performance Task *(continued)*

Part C

While the manager is tempted to set a new record by producing more units than have ever been produced in a year, he knows that people are not buying as many components as they used to, and he does not want to make more units than can be sold. He decides to set the goal of producing the third highest number of components in company history. Suggest a goal for the factory manager.

1,570,500

Part D

In researching the company financial reports, the factory manager discovers that the factory must produce at least 1,570,000 units in a year in order to make a profit. Does your goal from **Part C** meet this requirement? If so, explain why. If not, offer the factory manager a new goal that meets **both** requirements.

Yes, my answer is larger than 1,570,000 and is the third highest.

Part E

The factory manager's supervisor indicates that it is absolutely essential that the total number of units sold in years 6, 7, and the new year 8 be at least 4,500,000. Explain why the goal you gave the factory manager in **Part D** will also meet this new requirement.

The goal will meet this last requirement because it is large enough for the total to be greater than 4,500,000.

NAME

DATE

SCORE

Performance Task

Setting Goals

A factory produces electronic components. The new manager wants to set a goal for how many units will be produced in the upcoming year.

Write your answers on another piece of paper. Show all your work to receive full credit.

Part A

The factory has existed for seven years. The chart below gives the number of components produced by the factory each year.

Year	Components	
1	1,432,426	6
2	1,532,199	3
3	1,432,501	5
4	1,570,672	2
5	1,600,121	1
6	1,423,411	7
7	1,531,199	4

The factory manager needs to put the data in order so that he can make a decision on the next year's goal. Order the data from least to greatest.

Part B

The factory manger asks his assistant manager to give input for the production goal. The assistant manager suggests 1,423,000 units. Explain why this goal may not be appropriate.

This goal is lower than some years.

Performance Task *(continued)*

Part C

While the manager is tempted to set a new record by producing more units than have ever been produced in a year, he knows that people are not buying as many components as they used to, and he does not want to make more units than can be sold. He decides to set the goal of producing the third highest number of components in company history. Suggest a goal for the factory manager.

1,532,100

Part D

In researching the company financial reports, the factory manager discovers that the factory must produce at least 1,570,000 units in a year in order to make a profit. Does your goal from **Part C** meet this requirement? If so, explain why. If not, offer the factory manager a new goal that meets *both* requirements.

No, my answer does not meet the requirement.

Part E

The factory manager's supervisor indicates that it is absolutely essential that the total number of units sold in years 6, 7, and the new year 8 be at least 4,500,000. Explain why the goal you gave the factory manager in **Part D** will also meet this new requirement.

They add up to more than 4,500,000.

Student Model

NAME _____ DATE _____

SCORE _____

Performance Task

Setting Goals

A factory produces electronic components. The new manager wants to set a goal for how many units will be produced in the upcoming year.

Write your answers on another piece of paper. Show all your work to receive full credit.

Part A

The factory has existed for seven years. The chart below gives the number of components produced by the factory each year.

Year	Components	
1	1,432,426	1
2	1,532,199	4
3	1,432,501	3
4	1,570,672	6
5	1,600,121	7
6	1,423,411	2
7	1,531,199	5

The factory manager needs to put the data in order so that he can make a decision on the next year's goal. Order the data from least to greatest.

Part B

The factory manger asks his assistant manager to give input for the production goal. The assistant manager suggests 1,423,000 units. Explain why this goal may not be appropriate.

This goal is too high.

Performance Task *(continued)*

Part C

While the manager is tempted to set a new record by producing more units than have ever been produced in a year, he knows that people are not buying as many components as they used to, and he does not want to make more units than can be sold. He decides to set the goal of producing the third highest number of components in company history. Suggest a goal for the factory manager.

1,531,160

Part D

In researching the company financial reports, the factory manager discovers that the factory must produce at least 1,570,000 units in a year in order to make a profit. Does your goal from **Part C** meet this requirement? If so, explain why. If not, offer the factory manager a new goal that meets *both* requirements.

No. It is too low.

Part E

The factory manager's supervisor indicates that it is absolutely essential that the total number of units sold in years 6, 7, and the new year 8 be at least 4,500,000. Explain why the goal you gave the factory manager in **Part D** will also meet this new requirement.

It is big enough.

Page 127 • Buying Cards

Task Scenario
Students will multiply whole numbers and create a diagram to model a trading card purchase order.

Depth of Knowledge	DOK2, DOK3

Part	Maximum Points	Scoring Rubric
A	2	**Full Credit:** Sample answer: Her estimate is lower because she rounded down 12 to 10 and 24 to 20. Partial Credit (1 point) will be given for correctly saying her estimate will be lower without providing a reason. No credit will be given for an incorrect answer.
B	2	**Full Credit:** Sample answer: I would round 24 to 25 and round 12 to 10. This will be more accurate because I rounded the 24 to 25 which is closer than 20. Partial Credit (1 point) will be given for suggesting a more accurate way of rounding **OR** the giving an explanation as to why it would be more accurate. No credit will be given for an incorrect answer.
C	4	**Full Credit:** Sample answer: _288_ total packs Partial Credit (2 points) will be given for correctly labeling the diagram **OR** correctly identifying that there will be 288 total packs. No credit will be given for an incorrect answer.

Performance Task Rubrics

Part	Maximum Points	Scoring Rubric
D	2	Full Credit: 2304 cards; Sample answer: I found there were 288 packs in 12 boxes, so I multiplied that result by 8 cards per pack. Partial Credit (2 points) will be given for correctly identifying that there will be 288 cards **OR** giving an appropriate explanation of how to find the solution. No credit will be given for an incorrect answer.
TOTAL	10	

NAME _____   DATE _____

SCORE _____

Performance Task

Buying Cards

Keena is ordering baseball cards for her store. The boxes come with 24 packs and each pack has 8 cards.

Write your answers on another piece of paper. Show all your work to receive full credit.

Part A

A customer has requested 12 boxes. Keena estimates the order will be about 200 packs. Is her estimate higher or lower than the actual total? Explain your reasoning.

Her estimate is lower because she rounded down 12 to 10 and 24 to 20.

Part B

Suggest a more accurate way of estimating the number of total packs in 8 boxes. Explain why your estimate will be more accurate.

I would round 24 to 25 and round 12 to 10. This will be more accurate because I rounded the 24 to 25, which is closer than 20.

Performance Task *(continued)*

Part C

The store used the following area model to find the total number of packs in 12 boxes. Complete the labeling shown.

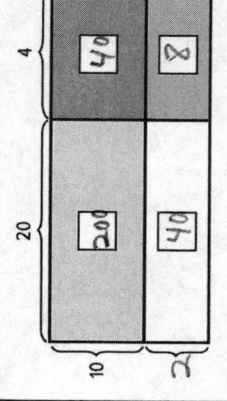

_288_ total packs

Part D

How many total cards are in 12 boxes? Explain your reasoning.

2304 cards

I found there were 288 packs in 12 boxes, so I multiplied that result by 8 cards per pack.

Student Model

NAME _____ DATE _____

SCORE _____

Performance Task

Buying Cards

Keena is ordering baseball cards for her store. The boxes come with 24 packs and each pack has 8 cards.

Write your answers on another piece of paper. Show all your work to receive full credit.

Part A

A customer has requested 12 boxes. Keena estimates the order will be about 200 packs. Is her estimate higher or lower than the actual total? Explain your reasoning.

Lower because she rounded down both numbers before multiplying.

Part B

Suggest a more accurate way of estimating the number of total packs in 8 boxes. Explain why your estimate will be more accurate.

Round 24 up to 25 and round 12 down to 10.

Performance Task (continued)

Part C

The store used the following area model to find the total number of packs in 12 boxes. Complete the labeling shown.

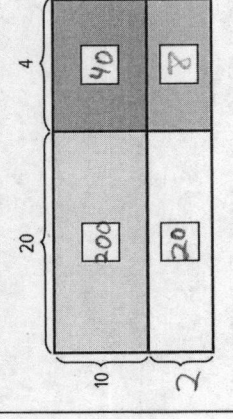

268 total packs

Part D

How many total cards are in 12 boxes? Explain your reasoning.

2144 ~ Multiply 268 by 8

NAME _____ DATE _____

SCORE _____

Performance Task

Buying Cards

Keena is ordering baseball cards for her store. The boxes come with 24 packs and each pack has 8 cards.

Write your answers on another piece of paper. Show all your work to receive full credit.

Part A

A customer has requested 12 boxes. Keena estimates the order will be about 200 packs. Is her estimate higher or lower than the actual total? Explain your reasoning.

lower

Part B

Suggest a more accurate way of estimating the number of total packs in 8 boxes. Explain why your estimate will be more accurate.

Round one number up and one number down

Performance Task (continued)

Part C

The store used the following area model to find the total number of packs in 12 boxes. Complete the labeling shown.

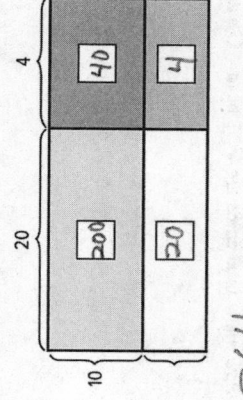

264 total packs

Part D

How many total cards are in 12 boxes? Explain your reasoning.

2,112 because there are 264 packs with 8 cards each

Student Model

NAME _____ DATE _____

Performance Task

SCORE _____

Buying Cards

Keena is ordering baseball cards for her store. The boxes come with 24 packs and each pack has 8 cards.

Write your answers on another piece of paper. Show all your work to receive full credit.

Part A

A customer has requested 12 boxes. Keena estimates the order will be about 200 packs. Is her estimate higher or lower than the actual total? Explain your reasoning.

higher

Part B

Suggest a more accurate way of estimating the number of total packs in 8 boxes. Explain why your estimate will be more accurate.

Multiply 25 by 10

Performance Task *(continued)*

Part C

The store used the following area model to find the total number of packs in 12 boxes. Complete the labeling shown.

_____ total packs

Part D

How many total cards are in 12 boxes? Explain your reasoning.

$$\begin{array}{r} 288 \\ \times\ 8 \\ \hline 2304 \end{array}$$

Page 129 • Saving for a Bike

Task Scenario
Students will use division by a single-digit number to help figure out how Janelle can earn enough money to buy a bike.

Depth of Knowledge	DOK2, DOK3	

Part	Maximum Points	Scoring Rubric
A	2	**Full Credit:** $81 per month; Sample Answer: I divided $486 ÷ 6, and got $81 Partial Credit (1 point) will be given for the correct answer without an explanation OR for a correct explanation with an incorrect answer. No credit will be given for an incorrect answer.
B	2	**Full Credit:** 18 cars; Sample answer: I added $81 and $81 to get her goal for two months, which is $162. Then I divided $162 ÷ $9 and got 18 cars Partial Credit (1 point) will be given for the correct answer without an explanation OR for a correct explanation with an incorrect answer. No credit will be given for an incorrect answer.
C	2	**Full Credit:** 21 lawns; Sample answer: I added $81 and $81 and got $162. Then I divided $162 ÷ $8 and got 20 R 2. If she wants to make her goal, she needs to mow an extra lawn because of the remainder. Partial Credit (1 point) will be given for the correct answer without an explanation or for forgetting to round up due to the remainder. No credit will be given for an incorrect answer.

Performance Task Rubrics

Part	Maximum Points	Scoring Rubric
D	3	**Full Credit:** Sample answer: She has $156 dollars left to go. I multiplied 21 by $8 and got $168. Then I multiplied 18 by $9 and got $162. Then I added $168 and $162 and got $330. Then I subtracted $486 − $330 and got $156 left. She needs to walk 32 dogs. I divided $156 ÷ $5 and got 31 R 1. Because she needs to make her goal, she needs to walk an extra dog. Partial Credit (2 points) will be given for obtaining one of the two answers with explanation. No credit will be given for two incorrect answers or one correct answer with no explanation.
TOTAL	9	

NAME _____ DATE _____

SCORE _____

Performance Task

Saving for a Bike

Janelle is trying to save money in order to purchase a bike. The bike costs $486. She has three different ways of making money, which are shown in the table below.

Mowing Lawns	$8 per lawn
Walking Dogs	$5 per walk
Washing Cars	$9 per car

Write your answers on another piece of paper. Show all your work to receive full credit.

Part A

Janelle wants to pay for the bike in 6 months. How much does she need to save each month in order to accomplish her goal? Explain.

$81/month

I divided 486 ÷ 6.

Performance Task *(continued)*

Part B

In the first two months, Janelle only washes cars. How many cars does she need to wash in order to make her goal for the first two months? Explain.

81 + 81 = 162   9)162
                 −9
                 72

18 cars

Part C

In the third and fourth months, Janelle only mows lawns. How many lawns does she need to mow in order to make her goal? Explain the meaning of the remainder.

        20 R2
8)162
 −16
   02

The remainder means she needs to do one more lawn → 21

Part D

How much does Janelle have left to earn? If she only walks dogs for the last two months, how many dogs will she need to walk to make her final goal for buying the bike? Explain

486 − 162 − 168 = 156

        31 R1
5)156
 −15
   06
   −5
    1

32 dogs

Student Model

NAME _____ DATE _____

SCORE _____

Performance Task

Saving for a Bike

Janelle is trying to save money in order to purchase a bike. The bike costs $486. She has three different ways of making money, which are shown in the table below.

Mowing Lawns	$8 per lawn
Walking Dogs	$5 per walk
Washing Cars	$9 per car

Write your answers on another piece of paper. Show all your work to receive full credit.

Part A

Janelle wants to pay for the bike in 6 months. How much does she need to save each month in order to accomplish her goal? Explain.

$486 \div 6 = 81$

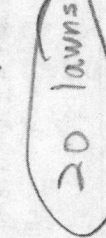

$81

Performance Task (continued)

Part B

In the first two months, Janelle only washes cars. How many cars does she need to wash in order to make her goal for the first two months? Explain.

$81 \div 9 = 9$

$9 + 9 = 18$ cars

Part C

In the third and fourth months, Janelle only mows lawns. How many lawns does she need to mow in order to make her goal? Explain the meaning of the remainder.

$81 + 81 = 162$ $162 \div 8 = 20\ R2$

20 lawns

Part D

How much does Janelle have left to earn? If she only walks dogs for the last two months, how many dogs will she need to walk to make her final goal for buying the bike? Explain

$486 - 162 - 162 = 162$

$162 \div 5 = 32\ R2$

32 dogs

NAME _____ DATE _____

SCORE _____

Performance Task

Saving for a Bike

Janelle is trying to save money in order to purchase a bike. The bike costs $486. She has three different ways of making money, which are shown in the table below.

Mowing Lawns	$8 per lawn
Walking Dogs	$5 per walk
Washing Cars	$9 per car

Write your answers on another piece of paper. Show all your work to receive full credit.

Part A

Janelle wants to pay for the bike in 6 months. How much does she need to save each month in order to accomplish her goal? Explain.

$81

Performance Task *(continued)*

Part B

In the first two months, Janelle only washes cars. How many cars does she need to wash in order to make her goal for the first two months? Explain.

9 + 9 = 18

Part C

In the third and fourth months, Janelle only mows lawns. How many lawns does she need to mow in order to make her goal? Explain the meaning of the remainder.

81 + 81 = 152

81) 152
  -8
  72

19 lawns

Part D

How much does Janelle have left to earn? If she only walks dogs for the last two months, how many dogs will she need to walk to make her final goal for buying the bike? Explain

152 + 162 = 314

39
5) 172
  -15
   22
   -20
    2

486 - 314 = 172

34 dogs

Student Model

NAME _____ DATE _____

SCORE _____

Performance Task

Saving for a Bike

Janelle is trying to save money in order to purchase a bike. The bike costs $486. She has three different ways of making money, which are shown in the table below.

Mowing Lawns	$8 per lawn
Walking Dogs	$5 per walk
Washing Cars	$9 per car

Write your answers on another piece of paper. Show all your work to receive full credit.

Part A

Janelle wants to pay for the bike in 6 months. How much does she need to save each month in order to accomplish her goal? Explain.

6)486
-48
  06

$74

Performance Task (continued)

Part B

In the first two months, Janelle only washes cars. How many cars does she need to wash in order to make her goal for the first two months? Explain.

71+71 = 142

9)142 15
  9
  52
 -45
   7

15 cars

Part C

In the third and fourth months, Janelle only mows lawns. How many lawns does she need to mow in order to make her goal? Explain the meaning of the remainder.

8)142
 -8
  62
 -62
   0

18 lawns

Part D

How much does Janelle have left to earn? If she only walks dogs for the last two months, how many dogs will she need to walk to make her final goal for buying the bike? Explain

486
-142
  40
  200

40 dogs

Page 131 • Constructing Frames for an Art Gallery

Task Scenario Students will use division by a multi-digit number to help figure out how to construct picture frames from boards of various lengths.		
Depth of Knowledge		DOK2, DOK3
Part	**Maximum Points**	**Scoring Rubric**
A	2	Full Credit: 52 inches; Sample answer: The area is length times width, so I divided 3,796 by 73 and got 52. Partial Credit (1 point) will be given for the correct answer without an explanation. No credit will be given for an incorrect answer.
B	2	Full Credit: Yes; Sample answer: The 5-foot board is 60 inches long, so it can be used for the width. The 8-foot board is 96 inches, so it can be used for the length. Partial Credit (1 point) will be given for the correct answer without an explanation. No credit will be given for an incorrect answer.
C	2	Full Credit: $96; Sample answer: If the frames are all the same price, divide $1,152 by 12 to get $96 for each frame. Partial Credit (1 point) will be given for the correct answer without an explanation. No credit will be given for an incorrect answer.

Performance Task Rubrics

Part	Maximum Points	Scoring Rubric
D	3	**Full Credit:**

Full Credit:

Board Length	Boards needed for 12 frames	Total inches left over
5 feet	24	192
8 feet	24	552

Sample answer: Each frame needs two 5-foot boards and two 8-foot boards to make all four sides. Each 5-foot board has 8 inches left over. Each 8-foot board has 23 inches left over. Partial Credit (1 point) will be given for obtaining the number of boards but not the remaining inches. An additional point will be given for the correct explanation.
No credit will be given for two incorrect answers or one correct answer with no explanation.

Part	Maximum Points	Scoring Rubric
TOTAL	9	

NAME _____

DATE _____

SCORE _____

Performance Task

Constructing Frames for an Art Gallery

A woodworker is making picture frames for some rather large paintings for a local art gallery. Each of the paintings has an area of 3,796 square inches. An example of the paintings is shown below:

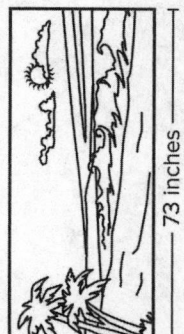

73 inches

Write your answers on another piece of paper. Show all your work to receive full credit.

Part A

Find the length of the painting? Explain.

$73 \times \ell = 3796$

$3796 \div 73 = 52$

Performance Task (continued)

Part B

The woodworker goes to a supply store and finds that boards are only sold in lengths of 5 feet and 8 feet. Are these boards long enough for the project? Explain.

5 feet = 60 inches > 52

8 feet = 96 inches > 73

Part C

The woodworker was contracted for 12 frames. He needs to make at least $1,152 in order to make the project worthwhile. How much does he need to charge for each frame?

$1152 \div 12 = 96$

$96

Part D

Fill in the chart for how many 5-foot boards and 8-foot boards the woodworker will need to fill the order, and fill in the amount of wood left over. Explain.

Board Length	Boards needed for 12 frames	Total inches left over
5 feet	24	192
8 feet	24	552

Each frame needs 2 of each board. 5 foot boards have 8 inches left over. 8 foot boards have 23.

Student Model

NAME _____

DATE _____

SCORE _____

Performance Task

Constructing Frames for an Art Gallery

A woodworker is making picture frames for some rather large paintings for a local art gallery. Each of the paintings has an area of 3,796 square inches. An example of the paintings is shown below:

73 inches

Write your answers on another piece of paper. Show all your work to receive full credit.

Part A

Find the length of the painting? Explain.

length x width = area, so divide the area by the width

$3796 \div 73 = 52$

Performance Task *(continued)*

Part B

The woodworker goes to a supply store and finds that boards are only sold in lengths of 5 feet and 8 feet. Are these boards long enough for the project? Explain.

Yes, they are longer than the sides.

Part C

The woodworker was contracted for 12 frames. He needs to make at least $1,152 in order to make the project worthwhile. How much does he need to charge for each frame?

Divide by 12

$96

Part D

Fill in the chart for how many 5-foot boards and 8-foot boards the woodworker will need to fill the order, and fill in the amount of wood left over. Explain.

Board Length	Boards needed for 12 frames	Total inches left over
5 feet	24	192
8 feet	24	552

NAME _____ DATE _____

SCORE _____

Performance Task

Constructing Frames for an Art Gallery

A woodworker is making picture frames for some rather large paintings for a local art gallery. Each of the paintings has an area of 3,796 square inches. An example of the paintings is shown below:

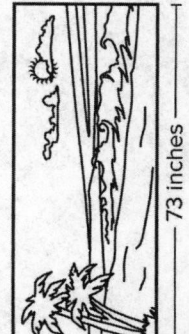

|← 73 inches →|

Write your answers on another piece of paper. Show all your work to receive full credit.

Part A

Find the length of the painting? Explain.

52 inches

Performance Task (continued)

Part B

The woodworker goes to a supply store and finds that boards are only sold in lengths of 5 feet and 8 feet. Are these boards long enough for the project? Explain.

Yes

Part C

The woodworker was contracted for 12 frames. He needs to make at least $1,152 in order to make the project worthwhile. How much does he need to charge for each frame?

$84

Part D

Fill in the chart for how many 5-foot boards and 8-foot boards the woodworker will need to fill the order, and fill in the amount of wood left over. Explain.

Board Length	Boards needed for 12 frames	Total inches left over
5 feet	24	180
8 feet	24	502

Student Model

NAME _____ DATE _____

SCORE _____

Performance Task

Constructing Frames for an Art Gallery

A woodworker is making picture frames for some rather large paintings for a local art gallery. Each of the paintings has an area of 3,796 square inches. An example of the paintings is shown below:

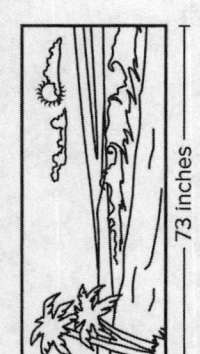

— 73 inches —

Write your answers on another piece of paper. Show all your work to receive full credit.

Part A

Find the length of the painting? Explain.

$$\begin{array}{r} 50 \\ 73\overline{)3796} \\ -365 \\ \hline 46 \end{array}$$

50 inches

Performance Task *(continued)*

Part B

The woodworker goes to a supply store and finds that boards are only sold in lengths of 5 feet and 8 feet. Are these boards long enough for the project? Explain.

5 feet > 50 inches

8 feet > 73 inches

Part C

The woodworker was contracted for 12 frames. He needs to make at least $1,152 in order to make the project worthwhile. How much does he need to charge for each frame?

$100

Part D

Fill in the chart for how many 5-foot boards and 8-foot boards the woodworker will need to fill the order, and fill in the amount of wood left over. Explain.

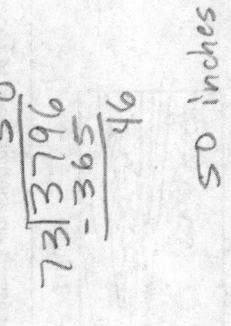

Board Length	Boards needed for 12 frames	Total inches left over
5 feet	12	120
8 feet	12	276

Page 133 • Planning for a Trip

Task Scenario
Students will use decimal addition, subtraction, and estimation to analyze a savings plan for an upcoming trip.

Depth of Knowledge	DOK2, DOK3

Part	Maximum Points	Scoring Rubric
A	2	Full Credit: 480 + 510 + 280 = 1,270 miles Partial Credit (1 point) will be given for a reasonable estimate without showing the individual terms. No credit will be given for an incorrect answer.
B	2	Full Credit: 51 gallons; $204 Partial Credit (1 point) will be given for having one of the two correct answers, based on the student's estimate in **Part A**. No credit will be given for an two incorrect answers.
C	2	Full Credit: $451.65; $18.15 + $22.62 + $35.67 + $129.00 + $18.15 + $22.62 + $35.67 + $129.00 + $18.15 + $22.62 Partial Credit (1 point) will be given for the correct answer without showing the expression **OR** for showing the correct expression without the correct answer. No credit will be given for an incorrect answer.

Performance Task Rubrics

Part	Maximum Points	Scoring Rubric
D	3	Full Credit: $44.35; Sample answer: First I added $451.65 + $204.00 and got $655.65. Then I subtracted $700.00 − $655.65 and got $44.35 left over Partial Credit (1 point) will be given for the correct answer without showing the calculation or for having one of the two correct steps and a calculation. No credit will be given for two incorrect answers or one correct answer with no explanations.
TOTAL	9	

NAME

DATE

SCORE

Performance Task

Planning for a Trip

The Perez family is planning for their summer vacation. The drive will take them three days, and the table shows how many miles they will drive each day.

Day 1	481.23 miles
Day 2	512.94 miles
Day 3	282.22 miles

Write your answers on another piece of paper. Show all your work to receive full credit.

Part A

Estimate the total number of miles that the Perez family will travel for the trip by rounding the number of miles each day to the nearest 10? Show your estimates.

```
  480
  510
  280
1,270
```

Part B

The family car will get 25 miles per gallon of gas. Use your estimate to determine how many gallons of gas the family will need to buy. Round to the nearest gallon. If gas is $4 per gallon, find out how much money they will need to budget for gas.

1,270 ÷ 25 → 51 gallons × $4

$204

Performance Task *(continued)*

Part C

Based on last year's trip, Mr. Sanchez has planned the following expenses for meals and hotel for one day.

Breakfast	$18.15	x3
Lunch	$22.62	x3
Dinner	$35.67	x2
Hotel	$129.00	x2

```
   54.45
   67.86
   71.34
+ 258.00
  451.65
```

The first two days, the family will eat all three meals and will need a hotel room. The third day they will need to eat only breakfast and lunch, and they will not need a hotel room. How much should the family expect to spend on food and lodging? Explain.

Part D

Mr. Sanchez has saved $700 for the trip. Given the estimate for the cost of gas and the expenses for food and lodging, how much can he expect to have left at the end of the three days. Explain.

```
  451.65
+ 204.00
  655.65
```

$$700 - 655.65 = 44.35$$

Student Model

Performance Task (continued)

Part C

Based on last year's trip, Mr. Sanchez has planned the following expenses for meals and hotel for one day.

Breakfast	$18.15	(3)
Lunch	$22.62	(3)
Dinner	$35.67	(2)
Hotel	$129.00	(2)

The first two days, the family will eat all three meals and will need a hotel room. The third day they will need to eat only breakfast and lunch, and they will not need a hotel room. How much should the family expect to spend on food and lodging? Explain.

$451.65   Multiply by number
of times and add.

Part D

Mr. Sanchez has saved $700 for the trip. Given the estimate for the cost of gas and the expenses for food and lodging, how much can he expect to have left at the end of the three days. Explain.

700 - 200 = 500

500 - 451.65 = 49.35

NAME .... DATE ....

SCORE ....

Performance Task

Planning for a Trip

The Perez family is planning for their summer vacation. The drive will take them three days, and the table shows how many miles they will drive each day.

Day 1	481.23 miles
Day 2	512.94 miles
Day 3	282.22 miles

Write your answers on another piece of paper. Show all your work to receive full credit.

Part A

Estimate the total number of miles that the Perez family will travel for the trip by rounding the number of miles each day to the nearest 10? Show your estimates.

480 + 510 + 280

1270

Part B

The family car will get 25 miles per gallon of gas. Use your estimate to determine how many gallons of gas the family will need to buy. Round to the nearest gallon. If gas is $4 per gallon, find out how much money they will need to budget for gas.

1270 ÷ 25 = 50.8

50 gallons $200

Performance Task (continued)

Part C

Based on last year's trip, Mr. Sanchez has planned the following expenses for meals and hotel for one day.

Breakfast	$18.15
Lunch	$22.62
Dinner	$35.67
Hotel	$129.00

The first two days, the family will eat all three meals and will need a hotel room. The third day they will need to eat only breakfast and lunch, and they will not need a hotel room. How much should the family expect to spend on food and lodging? Explain.

18.15   35.67
18.15   35.67
18.15   129.00
22.62   129.00
22.62   319.34
22.62
(22.13)

319.34
+122.3
441.65

Part D

Mr. Sanchez has saved $700 for the trip. Given the estimate for the cost of gas and the expenses for food and lodging, how much can he expect to have left at the end of the three days. Explain.

700.00
- 441.65
258.35

NAME _____ DATE _____

SCORE _____

Performance Task

Planning for a Trip

The Perez family is planning for their summer vacation. The drive will take them three days, and the table shows how many miles they will drive each day.

Day 1	481.23 miles
Day 2	512.94 miles
Day 3	282.22 miles

Write your answers on another piece of paper. Show all your work to receive full credit.

Part A

Estimate the total number of miles that the Perez family will travel for the trip by rounding the number of miles each day to the nearest 10? Show your estimates.

1,270

Part B

The family car will get 25 miles per gallon of gas. Use your estimate to determine how many gallons of gas the family will need to buy. Round to the nearest gallon. If gas is $4 per gallon, find out how much money they will need to budget for gas.

1270 ÷ 25 = 50
50 × $4 = $200

Student Model

Performance Task *(continued)*

Part C

Based on last year's trip, Mr. Sanchez has planned the following expenses for meals and hotel for one day.

Breakfast	$18.15
Lunch	$22.62
Dinner	$35.67
Hotel	$129.00

The first two days, the family will eat all three meals and will need a hotel room. The third day they will need to eat only breakfast and lunch, and they will not need a hotel room. How much should the family expect to spend on food and lodging? Explain.

$$18.15$$
$$22.62$$
$$35.67$$
$$129.00$$
$$205.44$$

$$205.44 \times 3 = 616.32$$

Part D

Mr. Sanchez has saved $700 for the trip. Given the estimate for the cost of gas and the expenses for food and lodging, how much can he expect to have left at the end of the three days. Explain.

$$700.00$$
$$-616.32$$
$$83.68$$

NAME

DATE

SCORE

Performance Task

Planning for a Trip

The Perez family is planning for their summer vacation. The drive will take them three days, and the table shows how many miles they will drive each day.

Day 1	481.23 miles
Day 2	512.94 miles
Day 3	282.22 miles

Write your answers on another piece of paper. Show all your work to receive full credit.

Part A

Estimate the total number of miles that the Perez family will travel for the trip by rounding the number of miles each day to the nearest 10? Show your estimates.

$$500 + 500 + 300 = 1300$$

Part B

The family car will get 25 miles per gallon of gas. Use your estimate to determine how many gallons of gas the family will need to buy. Round to the nearest gallon. If gas is $4 per gallon, find out how much money they will need to budget for gas.

$$1300 \div 25 = 52$$

round of 50 $50 \times 4 = \$200$

Page 135 • Making a Healthy Snack Mix

Task Scenario

Students will use decimal multiplication and division to create a snack mix and calculate nutritional information.

Depth of Knowledge	DOK2, DOK3	
Part	**Maximum Points**	**Scoring Rubric**
A	2	Full Credit: 44.55 grams; $(2 \times 14.25) + (1.5 \times 0.14) + (1.5 \times 10.56)$ Partial Credit (1 point) will be given for correct answer without a supporting expression. No credit will be given for an incorrect answer.
B	2	Full Credit: 5.345 grams; $(2 \times 0.001) + (1.5 \times 3.56) + (1.5 \times 0.002)$ Partial Credit (1 point) will be given for correct answer without a supporting expression. No credit will be given for an incorrect answer.
C	2	Full Credit: 705 calories; $(2 \times 168.54) + (1.5 \times 90) + (1.5 \times 155.28)$ Partial Credit (1 point) will be given for correct answer without a supporting expression. No credit will be given for an incorrect answer.
D	1	Full Credit: 5 servings No credit will be given for an incorrect answer.

Performance Task Rubrics

Part	Maximum Points	Scoring Rubric
E	3	Full Credit:

Full Credit:

Total Fat per Serving	8.91 grams
Salt (Sodium) per Serving	1.069 grams
Calories per Serving	141

Partial Credit (1 point each) will be given for each correct answer. No credit will be given for three incorrect responses.

| TOTAL | 10 | |

NAME _____ DATE _____

SCORE _____

Performance Task

Making a Healthy Snack Mix

Arundhati wants to make a snack mix with the following healthy ingredients. The total fat, salt, and calories are listed for a serving.

	Almonds	Raisins	Banana Chips
Total Fat	14.25 grams	0.14 grams	10.56 grams
Salt (Sodium)	0.001 grams	3.56 grams	0.002 grams
Calories	168.54	90	155.28

Write your answers on another piece of paper. Show all your work to receive full credit.

Part A

Arundhati will include 2 servings of almonds, 1.5 servings of raisins, and 1.5 servings of banana chips. Find the total fat that is in her mix. Show the expression you used to determine the answer.

$$2 \times 14.25 + 1.5 \times 0.14 + 1.5 \times 10.56$$

44.55 grams

Part B

Find the total salt (sodium) that is in Arundhati's snack mix. Show the expression you used to determine the answer.

$$2 \times 0.001 + 1.5 \times 3.56 + 1.5 \times 0.002$$

5.345 grams

Performance Task (continued)

Part C

Find the total calories that are in her mix. Show the expression you used to determine the answer.

$$2 \times 168.54 + 1.5 \times 90 + 1.5 \times 155.28$$

705 calories

Part D

If Arundhati has 2 servings of almonds, 1.5 servings of raisins, and 1.5 servings of banana chips, how many total servings does she have in her mix?

$$2 + 1.5 + 1.5$$

5 servings

Part E

Arundhati wants to calculate the total fat, salt, and calories per serving in her snack mix. Use your answers from Parts A, B, and C, together with the number of servings from Part D to fill in the following table.

Divide by 5

Total Fat per Serving	8.91 grams
Salt (Sodium) per Serving	1.069 grams
Calories per Serving	141

Student Model

NAME _____

DATE _____

SCORE _____

Performance Task

Making a Healthy Snack Mix

Arundhati wants to make a snack mix with the following healthy ingredients. The total fat, salt, and calories are listed for a serving.

	Almonds	Raisins	Banana Chips
Total Fat	14.25 grams	0.14 grams	10.56 grams
Salt (Sodium)	0.001 grams	3.56 grams	0.002 grams
Calories	168.54	90	155.28

Write your answers on another piece of paper. Show all your work to receive full credit.

Part A

Arundhati will include 2 servings of almonds, 1.5 servings of raisins, and 1.5 servings of banana chips. Find the total fat that is in her mix. Show the expression you used to determine the answer.

$(2 \times 14.25) + (1.5 \times 0.14) + (1.5 \times 10.56) =$

44.55

Part B

Find the total salt (sodium) that is in Arundhati's snack mix. Show the expression you used to determine the answer.

2×
1.5×
1.5×

5.345

Performance Task (continued)

Part C

Find the total calories that are in her mix. Show the expression you used to determine the answer.

2×
1.5×
1.5×

705

Part D

If Arundhati has 2 servings of almonds, 1.5 servings of raisins, and 1.5 servings of banana chips, how many total servings does she have in her mix?

5

Part E

Arundhati wants to calculate the total fat, salt, and calories per serving in her snack mix. Use your answers from **Parts A, B,** and **C,** together with the number of servings from **Part D** to fill in the following table.

Total Fat per Serving	8.91
Salt (Sodium) per Serving	1.069
Calories per Serving	141

NAME _____ DATE _____

SCORE _____

Performance Task

Making a Healthy Snack Mix

Arundhati wants to make a snack mix with the following healthy ingredients. The total fat, salt, and calories are listed for a serving.

	Almonds	Raisins	Banana Chips
Total Fat	14.25 grams	0.14 grams	10.56 grams
Salt (Sodium)	0.001 grams	3.56 grams	0.002 grams
Calories	168.54	90	155.28

Write your answers on another piece of paper. Show all your work to receive full credit.

Part A

Arundhati will include 2 servings of almonds, 1.5 servings of raisins, and 1.5 servings of banana chips. Find the total fat that is in her mix. Show the expression you used to determine the answer.

45.55 grams

Part B

Find the total salt (sodium) that is in Arundhati's snack mix. Show the expression you used to determine the answer.

5.3 grams

Performance Task *(continued)*

Part C

Find the total calories that are in her mix. Show the expression you used to determine the answer.

705 calories

Part D

If Arundhati has 2 servings of almonds, 1.5 servings of raisins, and 1.5 servings of banana chips, how many total servings does she have in her mix?

5 servings

Part E

Arundhati wants to calculate the total fat, salt, and calories per serving in her snack mix. Use your answers from Parts A, B, and C, together with the number of servings from Part D to fill in the following table.

Total Fat per Serving	9
Salt (Sodium) per Serving	1
Calories per Serving	141

Student Model

NAME _____ DATE _____

SCORE _____

Performance Task

Making a Healthy Snack Mix

Arundhati wants to make a snack mix with the following healthy ingredients. The total fat, salt, and calories are listed for a serving.

	Almonds	Raisins	Banana Chips
Total Fat	14.25 grams	0.14 grams	10.56 grams
Salt (Sodium)	0.001 grams	3.56 grams	0.002 grams
Calories	168.54	90	155.28

Write your answers on another piece of paper. Show all your work to receive full credit.

Part A

Arundhati will include 2 servings of almonds, 1.5 servings of raisins, and 1.5 servings of banana chips. Find the total fat that is in her mix. Show the expression you used to determine the answer.

45.1 g

Part B

Find the total salt (sodium) that is in Arundhati's snack mix. Show the expression you used to determine the answer.

5g

Performance Task (continued)

Part C

Find the total calories that are in her mix. Show the expression you used to determine the answer.

700 cal

Part D

If Arundhati has 2 servings of almonds, 1.5 servings of raisins, and 1.5 servings of banana chips, how many total servings does she have in her mix?

4

Part E

Arundhati wants to calculate the total fat, salt, and calories per serving in her snack mix. Use your answers from **Parts A, B, and C**, together with the number of servings from **Part D** to fill in the following table.

Total Fat per Serving	11.3
Salt (Sodium) per Serving	1.25
Calories per Serving	175

Page 137 • Planting Trees

Part	Maximum Points	Scoring Rubric
Task Scenario		Students will use graphs and patterns to compare two strategies for planting trees.
Depth of Knowledge		DOK2, DOK3, DOK4
A	2	Full Credit: Sample Answer: Kanona's pattern is a "multiply by 2" pattern. Latoya's pattern is a "add 3" pattern Partial Credit (1 point) will be given for a correct answer to one of the two patterns. No credit will be given for two incorrect answers.
B	3	Full Credit: 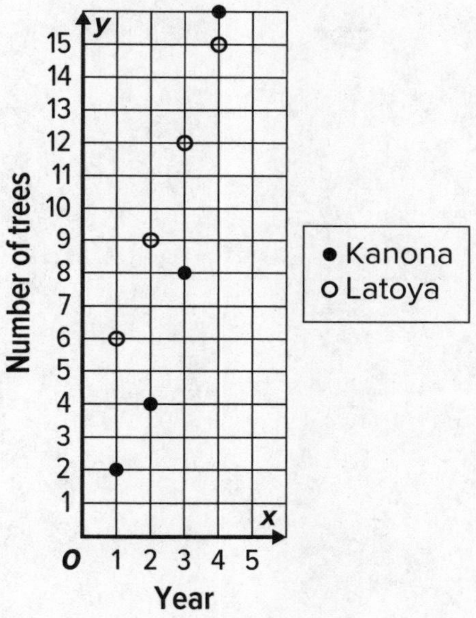 Partial Credit (2 points) will be given for having at least 6 of the 8 points correctly placed. Partial Credit (1 point) will be given for having at least 4 of the 8 points correctly placed. No credit will be given for fewer than four correct points.
C	1	Full Credit: Year 4 No credit will be given an incorrect answer.

Performance Task Rubrics

Part	Maximum Points	Scoring Rubric
D	2	Full Credit: | Year 1	8 trees
Year 2	13 trees
Year 3	20 trees
Year 4	31 trees	 Partial Credit (1 point) will be given for having at least two of the four correct answers. No credit will be given for fewer than 2 correct answers.
TOTAL	8	

NAME _____ DATE _____

SCORE _____

Performance Task

Planting Trees

Kanona and Latoya have each developed a plan for planting new trees in their local parks. Kanona will plant 2 trees the first year, 4 trees the second year, 8 trees the third year, and so on. Latoya will plant 6 trees the first year, 9 the second year, 12 the third year, and so on.

Write your answers on another piece of paper. Show all your work to receive full credit.

Part A

Describe the pattern for each girl's tree planting plan.

Kanona – multiply by 2

Latoya – add 3

Part B

For each of the first four years, plot the number of trees that each girl plans to plant. Use a different mark for each of the girls.

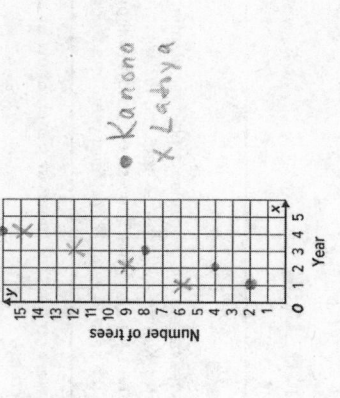

● Kanona
✗ Latoya

Performance Task *(continued)*

Part C

For the first couple of years Latoya plants more trees than Kanona. In what year will Kanona plant more trees than Latoya?

Year 4 (from the plot)

Part D

Fill in the table with the total number of trees planted in that year by both girls.

Year 1	8
Year 2	13
Year 3	20
Year 4	31

2+6
4+9
8+12
16+15

Student Model

NAME _____ DATE _____

SCORE _____

Performance Task

Planting Trees

Kanona and Latoya have each developed a plan for planting new trees in their local parks. Kanona will plant 2 trees the first year, 4 trees the second year, 8 trees the third year, and so on. Latoya will plant 6 trees the first year, 9 the second year, 12 the third year, and so on.

Write your answers on another piece of paper. Show all your work to receive full credit.

Part A

Describe the pattern for each girl's tree planting plan.

For Kanona, multiply by 2

For Latoya, add 3

Part B

For each of the first four years, plot the number of trees that each girl plans to plant. Use a different mark for each of the girls.

Number of trees — Year (graph with y-axis labeled 1–15, x-axis labeled 0 1 2 3 4 5)

Performance Task *(continued)*

Part C

For the first couple of years Latoya plants more trees than Kanona. In what year will Kanona plant more trees than Latoya?

Year 4

Part D

Fill in the table with the total number of trees planted in that year by both girls.

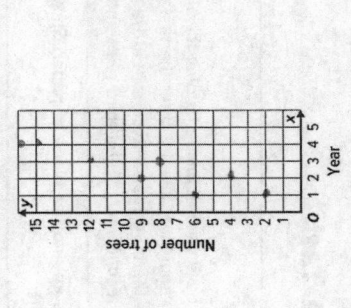

Year 1	8
Year 2	14
Year 3	20
Year 4	31

NAME _____ DATE _____

SCORE _____

Performance Task

Planting Trees

Kanona and Latoya have each developed a plan for planting new trees in their local parks. Kanona will plant 2 trees the first year, 4 trees the second year, 8 trees the third year, and so on. Latoya will plant 6 trees the first year, 9 the second year, 12 the third year, and so on.

Write your answers on another piece of paper. Show all your work to receive full credit.

Part A

Describe the pattern for each girl's tree planting plan.

K – add 2

L – add 3

Part B

For each of the first four years, plot the number of trees that each girl plans to plant. Use a different mark for each of the girls.

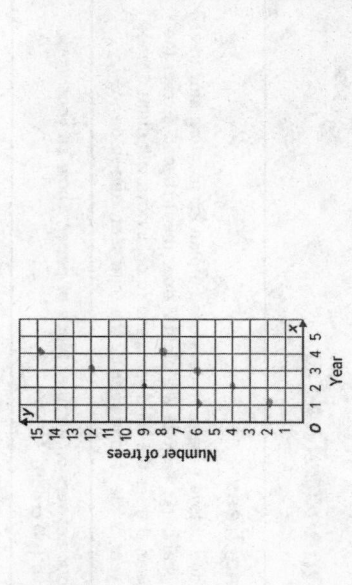

Performance Task *(continued)*

Part C

For the first couple of years Latoya plants more trees than Kanona. In what year will Kanona plant more trees than Latoya?

trick question!
It never happens

Part D

Fill in the table with the total number of trees planted in that year by both girls.

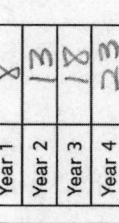

Year 1	8
Year 2	13
Year 3	18
Year 4	23

Student Model

NAME _____ DATE _____

SCORE _____

Performance Task

Planting Trees

Kanona and Latoya have each developed a plan for planting new trees in their local parks. Kanona will plant 2 trees the first year, 4 trees the second year, 8 trees the third year, and so on. Latoya will plant 6 trees the first year, 9 the second year, 12 the third year, and so on.

Write your answers on another piece of paper. Show all your work to receive full credit.

Part A

Describe the pattern for each girl's tree planting plan.

Add 2

Add 3

Part B

For each of the first four years, plot the number of trees that each girl plans to plant. Use a different mark for each of the girls.

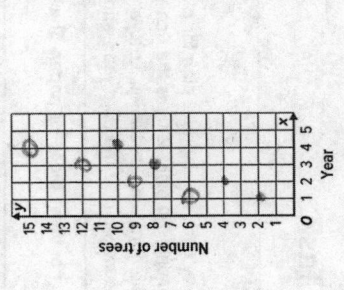

Performance Task *(continued)*

Part C

For the first couple of years Latoya plants more trees than Kanona. In what year will Kanona plant more trees than Latoya?

Year 3

Part D

Fill in the table with the total number of trees planted in that year by both girls.

Year 1	7
Year 2	12
Year 3	20
Year 4	26

Page 139 • Batting Averages

Task Scenario
Students will use fractions and decimals to calculate and compare batting averages.

Depth of Knowledge	DOK2, DOK3

Part	Maximum Points	Scoring Rubric
A	4	**Full Credit:**

Year	Fraction in lowest terms	Fraction as a decimal
1	$\frac{3}{10}$	0.3
2	$\frac{7}{20}$	0.35
3	$\frac{1}{5}$	0.2
4	$\frac{8}{25}$	0.32
5	$\frac{2}{5}$	0.4

Partial Credit (3 points) will be given for 6 or 7 correct answers, (2 points) for 4 or 5 correct answers, (1 point) for 2 or 3 correct answers.
No credit will be given for fewer than 2 correct answers.

Part	Maximum Points	Scoring Rubric
B	2	**Full Credit:** Year 5, Year 2, Year 4, Year 1, Year 3 Partial Credit (1 point) will be given for having the years in the reverse order. No credit will be given for any other ordering.
C	2	**Full Credit:** 100 at bats; 32 hits; 0.32 Partial Credit (1 point) will be given for a correct decimal but incorrect hits and/or at bats **OR** correct hits and at bats but incorrect decimal. No credit will be given an incorrectly calculated decimal.

Performance Task Rubrics

Part	Maximum Points	Scoring Rubric
D	2	Full Credit: 16 hits; $\frac{16}{50} = 0.32$ Partial Credit (1 point) will be given for a correct answer without an explanation **OR** for a correct explanation of how to find the number of hits but an incorrect number of hits. No credit will be given for an incorrect answer.
TOTAL	10	

Performance Task *(continued)*

Part B

Put each year in order from Winston's best performance to his worst.

Year 5
Year 2
Year 4
Year 1
Year 3

Part C

How many at bats did Winston have over the course of the five years? How many hits did he get in that time? Use this to find the decimal that represents the fraction of the time that Winston got a hit in the five-year period.

$3+7+4+8+10 = 32$ hits
$10+20+20+25+25 = 100$ at bats
$\frac{32}{100} = 0.32$

Part D

Winston's friend Vince had the exact same batting average in five years. However, Vince only batted 50 times. How many hits did Vince get in this time? Explain.

$\frac{32}{100} = \frac{16}{50}$

Performance Task

Batting Averages

Winston has been batting for the same baseball team for five years. The table shows how many times he was at bat and how many hits he got.

Year	At Bats	Hits
1	10	3
2	20	7
3	20	4
4	25	8
5	25	10

Write your answers on another piece of paper. Show all your work to receive full credit.

Part A

For each year, what fraction of the times that Winston went to bat did he get a hit? Write your answer as a fraction in lowest terms and as a decimal. The first year is done for you.

Year	Fraction in lowest terms	Fraction as a decimal
1	$\frac{3}{10}$	0.3
2	$\frac{7}{20}$	0.35
3	$\frac{1}{5}$	0.2
4	$\frac{8}{25}$	0.32
5	$\frac{2}{5}$	0.4

Student Model

Performance Task *(continued)*

Part B

Put each year in order from Winston's best performance to his worst.

3, 1, 4, 2, 5

Part C

How many at bats did Winston have over the course of the five years? How many hits did he get in that time? Use this to find the decimal that represents the fraction of the time that Winston got a hit in the five-year period.

$\frac{32}{100}$

Part D

Winston's friend Vince had the exact same batting average in five years. However, Vince only batted 50 times. How many hits did Vince get in this time? Explain.

$\frac{32}{100} \div 2 = \frac{16}{50}$

NAME _____ DATE _____

SCORE _____

Performance Task

Batting Averages

Winston has been batting for the same baseball team for five years. The table shows how many times he was at bat and how many hits he got.

Year	At Bats	Hits
1	10	3
2	20	7
3	20	4
4	25	8
5	25	10

100   32

Write your answers on another piece of paper. Show all your work to receive full credit.

Part A

For each year, what fraction of the times that Winston went to bat did he get a hit? Write your answer as a fraction in lowest terms and as a decimal. The first year is done for you.

Year	Fraction in lowest terms	Fraction as a decimal
1	$\frac{3}{10}$	0.3
2	$\frac{7}{20}$	0.35
3	$\frac{1}{4}$	0.25
4	$\frac{8}{25}$	0.32
5	$\frac{5}{12}$	0.4

NAME

DATE

SCORE

Performance Task

Batting Averages

Winston has been batting for the same baseball team for five years. The table shows how many times he was at bat and how many hits he got

Year	At Bats	Hits
1	10	3
2	20	7
3	20	4
4	25	8
5	25	10

Write your answers on another piece of paper. Show all your work to receive full credit.

Part A

For each year, what fraction of the times that Winston went to bat did he get a hit? Write your answer as a fraction in lowest terms and as a decimal. The first year is done for you.

Year	Fraction in lowest terms	Fraction as a decimal
1	$\frac{3}{10}$	0.3
2	$\frac{7}{20}$	0.35
3	$\frac{1}{7}$	0.25
4	$\frac{1}{3}$	0.33
5	$\frac{2}{6}$	0.4

Performance Task (continued)

Part B

Put each year in order from Winston's best performance to his worst.

Y5 > Y2 > Y4 > Y1 > Y3

Part C

How many at bats did Winston have over the course of the five years? How many hits did he get in that time? Use this to find the decimal that represents the fraction of the time that Winston got a hit in the five-year period.

34 hits

100 at bats

0.34

Part D

Winston's friend Vince had the exact same batting average in five years. However, Vince only batted 50 times. How many hits did Vince get in this time? Explain.

17 hits

Student Model

NAME _____ DATE _____

SCORE _____

Performance Task

Batting Averages

Winston has been batting for the same baseball team for five years. The table shows how many times he was at bat and how many hits he got.

Year	At Bats	Hits
1	10	3
2	20	7
3	20	4
4	25	8
5	25	10

Write your answers on another piece of paper. Show all your work to receive full credit.

Part A

For each year, what fraction of the times that Winston went to bat did he get a hit? Write your answer as a fraction in lowest terms and as a decimal. The first year is done for you.

Year	Fraction in lowest terms	Fraction as a decimal
1	$\frac{3}{10}$	0.3
2	$\frac{7}{20}$	0.33
3	$\frac{1}{5}$	0.2
4	$\frac{8}{25}$	0.33
5	$\frac{2}{5}$	0.4

Performance Task (continued)

Part B

Put each year in order from Winston's best performance to his worst.

year 5, year 2, year 4, year 1, year 3

Part C

How many at bats did Winston have over the course of the five years? How many hits did he get in that time? Use this to find the decimal that represents the fraction of the time that Winston got a hit in the five-year period.

29 hits
90 at bats
0.32

Part D

Winston's friend Vince had the exact same batting average in five years. However, Vince only batted 50 times. How many hits did Vince get in this time? Explain.

about 14

Page 141 • Triathlon Training

Task Scenario
Students will use addition and subtraction of fractions in order to plan a training schedule for two athletes preparing for a triathlon.

Depth of Knowledge	DOK2, DOK3	
Part	**Maximum Points**	**Scoring Rubric**
A	3	Full Credit: Minh; Sample answer: Minh swam $\frac{1}{3} + \frac{1}{2} + \frac{3}{4} + 1 + \frac{1}{4} = 2\frac{5}{6}$ miles. PJ swam $\frac{1}{4} + \frac{1}{2} + 1 + \frac{2}{3} + \frac{1}{4} = 2\frac{2}{3}$ miles. Partial Credit (1 point) will be given for Minh's total. Another (1 point) will be given for PJ's total. No credit will be given for two incorrect totals.
B	2	Full Credit: $11\frac{3}{4}$ miles; Sample answer: First I added $10\frac{1}{2} + 13\frac{1}{3} + 11\frac{3}{4} + 12\frac{2}{3}$ and got $48\frac{1}{4}$. Then I subtracted $60 - 48\frac{1}{4}$ and got $11\frac{3}{4}$. Partial Credit (1 point) will be given for the correct total with an incorrect difference. No credit will be given for having both an incorrect total and an incorrect difference.
C	3	Full Credit: PJ; Sample answer: For Minh, I subtracted $125\frac{1}{2} - 125\frac{1}{6}$ and got $\frac{1}{3}$. For PJ, I subtracted $120\frac{1}{4} - 120$ and got $\frac{1}{4}$. Then I compared, and $\frac{1}{3} > \frac{1}{4}$, so PJ was closer. Partial Credit (1 point) will be given for the correct difference for Minh, and (1 point) for the correct difference for PJ. No credit will be given for three incorrect answers.
TOTAL	8	

NAME _____

DATE _____

SCORE _____

Performance Task

Triathlon Training

Minh and PJ are training for a triathlon that involves swimming, biking, and running. They spend a week training for each event.

Write your answers on another piece of paper. Show all your work to receive full credit.

Part A

In Week 1 the two athletes are concentrating on swimming. The table shows how many miles each person swam on the given day.

Day	Minh	PJ
Monday	$\frac{1}{3}$ mile	$\frac{1}{4}$ mile
Tuesday	$\frac{1}{2}$ mile	$\frac{1}{2}$ mile
Wednesday	$\frac{3}{4}$ mile	1 mile
Thursday	1 mile	$\frac{2}{3}$ mile
Friday	$\frac{1}{4}$ mile	$\frac{1}{4}$ mile

Which athlete swam further? Explain.

Handwritten:

Minh (circled)

Minh: $\frac{1}{3} + \frac{1}{2} + \frac{3}{4} + 1 + \frac{1}{4} = 2\frac{5}{6}$

PJ: $\frac{1}{4} + \frac{1}{2} + 1 + \frac{2}{3} + \frac{1}{4} = 2\frac{3}{4}$

Performance Task *(continued)*

Part B

In Week 2 the two athletes are concentrating on running and decide to train together. Their goal is to run 60 miles in five days. On Monday they run $10\frac{1}{2}$ miles. On Tuesday they run $13\frac{1}{3}$ miles. On Wednesday they run $11\frac{3}{4}$ miles. On Thursday they run $12\frac{2}{3}$ miles. How far do they have to run on Friday to meet their goal? Explain.

Handwritten:

$10\frac{1}{2} + 13\frac{1}{3} + 11\frac{3}{4} + 12\frac{2}{3} = 48\frac{1}{4}$

$60 - 48\frac{1}{4} = 11\frac{3}{4}$

Part C

In Week 3 the two athletes concentrate on biking. Minh's goal is to bike $125\frac{1}{2}$ miles. At the end of the week, he found that he actually biked $125\frac{1}{6}$ miles. PJ's goal was to bike $120\frac{1}{4}$ miles. At the end of the week he found that he actually biked 120 miles. Which athlete was closer to making his goal? Explain.

Handwritten:

Minh: $125\frac{1}{2} - 125\frac{1}{6} = \frac{2}{6} = \frac{1}{3}$

PJ: $120\frac{1}{4} - 120 = \frac{1}{4}$

$\frac{1}{4} < \frac{1}{3}$

PJ (circled)

NAME _____

DATE _____

SCORE _____

Performance Task

Triathlon Training

Minh and PJ are training for a triathlon that involves swimming, biking, and running. They spend a week training for each event.

Write your answers on another piece of paper. Show all your work to receive full credit.

Part A

In Week 1 the two athletes are concentrating on swimming. The table shows how many miles each person swam on the given day.

Day	Minh	PJ
Monday	$\frac{1}{3}$ mile	$\frac{1}{4}$ mile
Tuesday	$\frac{1}{2}$ mile	$\frac{1}{2}$ mile
Wednesday	$\frac{3}{4}$ mile	1 mile
Thursday	1 mile	$\frac{2}{3}$ mile
Friday	$\frac{1}{4}$ mile	$\frac{1}{4}$ mile

Which athlete swam further? Explain.

$$\frac{1}{3} + \frac{1}{2} + \frac{3}{4} + 1 + \frac{1}{4} = 2\frac{9}{12}$$

$$\frac{1}{4} + \frac{1}{2} + 1 + \frac{2}{3} + \frac{1}{4} = 2\frac{8}{12}$$

PJ swam further.

Performance Task (continued)

Part B

In Week 2 the two athletes are concentrating on running and decide to train together. Their goal is to run 60 miles in five days. On Monday they run $10\frac{1}{2}$ miles. On Tuesday they run $13\frac{1}{3}$ miles. On Wednesday they run $11\frac{3}{4}$ miles. On Thursday they run $12\frac{2}{3}$ miles. How far do they have to run on Friday to meet their goal? Explain.

$$10\frac{1}{2} + 13\frac{1}{3} + 11\frac{3}{4} + 12\frac{2}{3} = 48\frac{3}{4}$$

$$60 - 48\frac{3}{4} = 11\frac{1}{4}$$

Part C

In Week 3 the two athletes concentrate on biking. Minh's goal is to bike $125\frac{1}{2}$ miles. At the end of the week, he found that he actually biked $125\frac{1}{6}$ miles. PJ's goal was to bike $120\frac{1}{4}$ miles. At the end of the week he found that he actually biked 120 miles. Which athlete was closer to making his goal? Explain.

$$125\frac{1}{2}$$
$$125\frac{1}{6}$$

$$\frac{1}{2} - \frac{1}{6} = \frac{3}{6} - \frac{1}{6} = \frac{2}{6} = \frac{1}{3}$$

$$120\frac{1}{4}$$
$$120$$

$$\frac{1}{4}$$

$$\frac{1}{3} < \frac{1}{4}$$

Minh was closer.

Student Model

NAME _____

DATE _____

SCORE _____

Performance Task

Triathlon Training

Minh and PJ are training for a triathlon that involves swimming, biking, and running. They spend a week training for each event.

Write your answers on another piece of paper. Show all your work to receive full credit.

Part A

In Week 1 the two athletes are concentrating on swimming. The table shows how many miles each person swam on the given day.

Day	Minh	PJ
Monday	$\frac{1}{3}$ mile	$\frac{1}{4}$ mile
Tuesday	$\frac{1}{2}$ mile	$\frac{1}{2}$ mile
Wednesday	$\frac{3}{4}$ mile	1 mile
Thursday	1 mile	$\frac{2}{3}$ mile
Friday	$\frac{1}{4}$ mile	$\frac{1}{4}$ mile

Which athlete swam further? Explain.

Minh - about 3 miles

PJ - about 2½ miles

Performance Task (continued)

Part B

In Week 2 the two athletes are concentrating on running and decide to train together. Their goal is to run 60 miles in five days. On Monday they run $10\frac{1}{2}$ miles. On Tuesday they run $13\frac{1}{3}$ miles. On Wednesday they run $11\frac{3}{4}$ miles. On Thursday they run $12\frac{2}{3}$ miles. How far do they have to run on Friday to meet their goal? Explain.

$10 + 13 + 12 + 11 = 46$

$\frac{1}{2} + \frac{1}{3} + \frac{3}{4} + \frac{2}{3} = 2$

$\overline{48}$

$60 - 48 = 12$

Part C

In Week 3 the two athletes concentrate on biking. Minh's goal is to bike $125\frac{1}{2}$ miles. At the end of the week, he found that he actually biked $125\frac{1}{6}$ miles. PJ's goal was to bike $120\frac{1}{4}$ miles. At the end of the week he found that he actually biked 120 miles. Which athlete was closer to making his goal? Explain.

$125\frac{1}{2}$
-125
$\overline{\frac{1}{2}}$

$120\frac{1}{4}$
-120
$\overline{\frac{1}{4}}$

$\frac{1}{2} > \frac{1}{4}$, so PJ was closer.

NAME

DATE

SCORE

Performance Task

Triathlon Training

Minh and PJ are training for a triathlon that involves swimming, biking, and running. They spend a week training for each event.

Write your answers on another piece of paper. Show all your work to receive full credit.

Part A

In Week 1 the two athletes are concentrating on swimming. The table shows how many miles each person swam on the given day.

Day	Minh	PJ
Monday	$\frac{1}{3}$ mile	$\frac{1}{4}$ mile
Tuesday	$\frac{1}{2}$ mile	$\frac{1}{2}$ mile
Wednesday	$\frac{3}{4}$ mile	1 mile
Thursday	1 mile	$\frac{2}{3}$ mile
Friday	$\frac{1}{4}$ mile	$\frac{1}{4}$ mile

Which athlete swam further? Explain.

Minh. Add the numbers.

Performance Task (continued)

Part B

In Week 2 the two athletes are concentrating on running and decide to train together. Their goal is to run 60 miles in five days. On Monday they run $10\frac{1}{2}$ miles. On Tuesday they run $13\frac{1}{3}$ miles. On Wednesday they run $11\frac{3}{4}$ miles. On Thursday they run $12\frac{2}{3}$ miles. How far do they have to run on Friday to meet their goal? Explain.

$$10\frac{1}{2} + 13\frac{1}{3} + 11\frac{3}{4} + 12\frac{2}{3}$$

They ran about 50 miles

Part C

In Week 3 the two athletes concentrate on biking. Minh's goal is to bike $125\frac{1}{2}$ miles. At the end of the week, he found that he actually biked $125\frac{1}{6}$ miles. PJ's goal was to bike $120\frac{1}{4}$ miles. At the end of the week he found that he actually biked 120 miles. Which athlete was closer to making his goal? Explain.

Minh was closer.

Student Model

Page 143 • Creating a Floor Plan

Task Scenario		
Students will use multiplication of fractions and modeling to create a floor plan and calculate the cost of constructing a new kitchen.		
Depth of Knowledge	DOK3, DOK4	

Part	Maximum Points	Scoring Rubric
A	4	Full Credit: Sample answer: I shaded $\frac{2}{3}$ of the half left for the kitchen and the living room. It is $\frac{1}{3}$ of the whole. The number sentence is $\frac{2}{3} \times \frac{1}{2} = \frac{1}{3}$. Partial Credit (2 points) will be given for a correct diagram without a number sentence. An additional point will be given for the answer of $\frac{1}{3}$ without a number sentence. No credit will be given for an incorrect answer.
B	2	Full Credit: $266\frac{2}{3}$ square feet; Sample answer: The area of the whole is $40 \times 20 = 800$ square feet. I multiplied 800 by $\frac{1}{3}$ and got $266\frac{2}{3}$ square feet. Partial Credit (1 point) will be given for a correct answer without an explanation. No credit will be given for an incorrect answer.

Part	Maximum Points	Scoring Rubric
C	2	Full Credit: $30,400; Sample answer: I multiplied $266\frac{2}{3}$ by \$114 and got \$30,400. Partial Credit (1 point) will be given for a correct answer without an explanation. No credit will be given for an incorrect answer.
TOTAL	8	

Performance Task Rubrics

NAME _____ DATE _____

SCORE _____

Performance Task

Creating a Floor Plan

An architect is trying to figure out how to layout a kitchen, a family room, and a living room in a new home. He wants $\frac{1}{2}$ of the floor to be the family room. Once that is marked off, the architect wants $\frac{2}{3}$ of the remaining space to be the kitchen and the rest to be the living room.

Write your answers on another piece of paper. Show all your work to receive full credit.

Part A

A model of the floor is shown below. Divide the floor into areas to find the fraction of the whole that the kitchen will take up. Shade the kitchen. Explain your answer and diagram with a number sentence.

$$\frac{2}{3} \times \frac{1}{2} = \frac{1}{3} \text{ of the whole area}$$

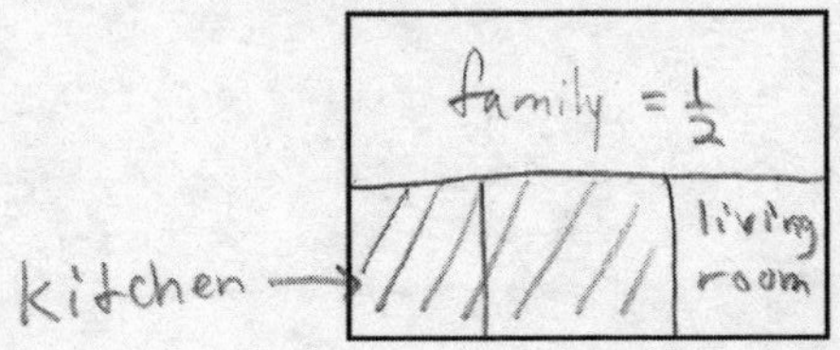

family = $\frac{1}{2}$

living room

kitchen →

Performance Task *(continued)*

Part B

The large rectangle is 40 feet by 20 feet. Find the area of the kitchen in square feet. Explain.

$$\frac{1}{3} \times 20 \times \frac{2}{3} \times 40 = 266\frac{2}{3} \text{ square feet}$$

Part C

The cost of building the kitchen will be $114 per square foot. Find the cost of constructing the kitchen. Explain.

$$266\frac{2}{3} \times 114 = \$30,400$$

NAME

DATE

SCORE

Performance Task

Creating a Floor Plan

An architect is trying to figure out how to layout a kitchen, a family room, and a living room in a new home. He wants $\frac{1}{2}$ of the floor to be the family room. Once that is marked off, the architect wants $\frac{2}{3}$ of the remaining space to be the kitchen and the rest to be the living room.

Write your answers on another piece of paper. Show all your work to receive full credit.

Part A

A model of the floor is shown below. Divide the floor into areas to find the fraction of the whole that the kitchen will take up. Shade the kitchen. Explain your answer and diagram with a number sentence.

family room $\frac{1}{2} = \frac{6}{12}$

kitchen $\frac{2}{3} \times \frac{1}{2} = \frac{1}{3}$

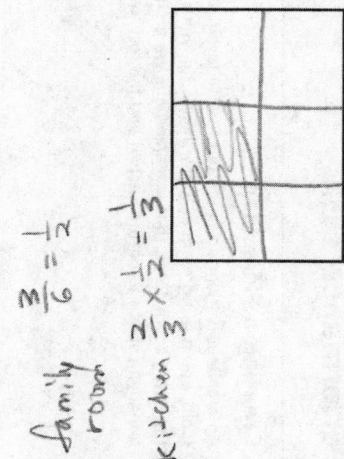

Performance Task (continued)

Part B

The large rectangle is 40 feet by 20 feet. Find the area of the kitchen in square feet. Explain.

Whole area = $40 \times 20 = 800$

Kitchen area = $\frac{1}{3} \times 800$

$= 300$

Part C

The cost of building the kitchen will be $114 per square foot. Find the cost of constructing the kitchen. Explain.

$\begin{array}{r} 114 \\ \times\ 300 \\ \hline 34200 \end{array}$

Student Model

NAME _____ DATE _____

SCORE _____

Performance Task

Creating a Floor Plan

An architect is trying to figure out how to layout a kitchen, a family room, and a living room in a new home. He wants $\frac{1}{2}$ of the floor to be the family room. Once that is marked off, the architect wants $\frac{2}{3}$ of the remaining space to be the kitchen and the rest to be the living room.

Write your answers on another piece of paper. Show all your work to receive full credit.

Part A

A model of the floor is shown below. Divide the floor into areas to find the fraction of the whole that the kitchen will take up. Shade the kitchen. Explain your answer and diagram with a number sentence.

$$\frac{2}{3} \times \frac{1}{2} = \frac{2}{6}$$

$$\boxed{\frac{1}{3}}$$

Performance Task (continued)

Part B

The large rectangle is 40 feet by 20 feet. Find the area of the kitchen in square feet. Explain.

$$800$$

$$800 \times \frac{2}{3} = 533\frac{1}{3} \text{ feet}$$

Part C

The cost of building the kitchen will be $114 per square foot. Find the cost of constructing the kitchen. Explain.

Multiply by Part B

$60,800

Performance Task *(continued)*

Part B

The large rectangle is 40 feet by 20 feet. Find the area of the kitchen in square feet. Explain.

$$533\frac{1}{3}$$

Part C

The cost of building the kitchen will be $114 per square foot. Find the cost of constructing the kitchen. Explain.

$$533 \times 114 = 60,762$$

Student Model

NAME _____ DATE _____

SCORE _____

Performance Task

Creating a Floor Plan

An architect is trying to figure out how to layout a kitchen, a family room, and a living room in a new home. He wants $\frac{1}{2}$ of the floor to be the family room. Once that is marked off, the architect wants $\frac{2}{3}$ of the remaining space to be the kitchen and the rest to be the living room.

Write your answers on another piece of paper. Show all your work to receive full credit.

Part A

A model of the floor is shown below. Divide the floor into areas to find the fraction of the whole that the kitchen will take up. Shade the kitchen. Explain your answer and diagram with a number sentence.

$$\frac{2}{3}$$

Page 145 • Comparing Mountains

Task Scenario
Students will use unit conversions in height, weight, and volume to compare mountains and prepare for a climb.

Depth of Knowledge	DOK2, DOK3

Part	Maximum Points	Scoring Rubric
A	3	**Full Credit:** 	Mountain	Height in Feet	Height in Miles and Feet	 |---|---|---| | Mt. Everest	29,029 feet	5 miles, 2,629 feet	 | Mt. McKinley	20,322 feet	3 miles, 4,482 feet	 | Mt. Kilimanjaro	19,341 feet	3 miles, 3,501 feet	 Partial Credit (1 point) will be given for each correct answer in miles. No credit will be given for three incorrect answers.
B	3	**Full Credit:** 	Mountain	Height in Meters	Height in kilometers	 |---|---|---| | Mt. Everest	8,848 meters	8.848 km	 | Mt. McKinley	6,194 meters	6.194 km	 | Mt. Kilimanjaro	5,895 meters	5.895 km	 Partial Credit (1 point) will be given for each correct answer in km. No credit will be given for three incorrect answers.
C	2	**Full Credit:** 320 ounces; Sample answer: $20 \times 16 = 320$ Partial Credit (1 point) will be given for the correct answer with no supporting calculation. No credit will be given for an incorrect answer.

Part	Maximum Points	Scoring Rubric
D	2	Full Credit: 1.5 liters; Sample answer: 1,500 ÷ 1,000 = 1.5 Partial Credit (1 point) will be given for the correct answer with no supporting calculation. No credit will be given for an incorrect answer.
TOTAL	10	

Performance Task Rubrics

NAME _____ DATE _____ SCORE _____

Performance Task

Comparing Mountains

A scientist wants to compare information on three different mountain hikes. Research ahead of time the height in feet of Mount Everest, Mount McKinley, and Mount Kilimanjaro.

Write your answers on another piece of paper. Show all your work to receive full credit.

Part A

From your research, fill in the heights in feet of the three mountains. Then convert the measurements to miles and feet.

Mountain	Height in Feet	Height in Miles and Feet
Mt. Everest	29,029	5 miles, 2,629 ft
Mt. McKinley	20,322	3 miles, 4,482 ft
Mt Kilimanjaro	19,341	3 miles, 3,501 ft

Performance Task (continued)

Part B

From your research, fill in the heights in meters of the three mountains. Then convert the measurements to kilometers.

Mountain	Height in Meters	Height in kilometers
Mt. Everest	8,848	8,848
Mt. McKinley	6,194	6,194
Mt. Kilimanjaro	5,895	5,895

Part C

The weight of a day pack containing hiking supplies is 20 pounds. How many ounces is this? Show your calculation.

$20 \times 16 = 320$ ounces

Part D

Each day before climbing, it is recommended that a climber drink 1,500 milliliters of water. How many liters is this? Explain.

$1,500 \div 1,000 = 1.5 L$

NAME _____

DATE _____

SCORE _____

Performance Task

Comparing Mountains

A scientist wants to compare information on three different mountain hikes. Research ahead of time the height in feet of Mount Everest, Mount McKinley, and Mount Kilimanjaro.

Write your answers on another piece of paper. Show all your work to receive full credit.

Part A

From your research, fill in the heights in feet of the three mountains. Then convert the measurements to miles and feet.

Mountain	Height in Feet	Height in Miles and Feet
Mt. Everest	29,029	5 miles ft
Mt McKinley	20,322	3 miles 4,482 ft
Mt Kilimanjaro	19,341	3 miles 3,501 ft

Performance Task (continued)

Part B

From your research, fill in the heights in meters of the three mountains. Then convert the measurements to kilometers.

Mountain	Height in Meters	Height in kilometers
Mt. Everest	8,848	8.848
Mt McKinley	6,194	6.19
Mt Kilimanjaro	5,895	5.895

Part C

The weight of a day pack containing hiking supplies is 20 pounds. How many ounces is this? Show your calculation.

$$16$$
$$\times 20$$
$$320$$

Part D

Each day before climbing, it is recommended that a climber drink 1,500 milliliters of water. How many liters is this? Explain.

$$1{,}500 \div 1{,}000$$

$$1.5$$

Student Model

NAME

DATE

Performance Task

Comparing Mountains

A scientist wants to compare information on three different mountain hikes. Research ahead of time the height in feet of Mount Everest, Mount McKinley, and Mount Kilimanjaro.

Write your answers on another piece of paper. Show all your work to receive full credit.

Part A

From your research, fill in the heights in feet of the three mountains. Then convert the measurements to miles and feet.

Mountain	Height in Feet	Height in Miles and Feet
Mt. Everest	30,000	5 miles 3,600
Mt. McKinley	20,322	3 miles
Mt. Kilimanjaro	19,341	3 miles 350

Performance Task (continued)

Part B

From your research, fill in the heights in meters of the three mountains. Then convert the measurements to kilometers.

Mountain	Height in Meters	Height in kilometers
Mt. Everest	9,000	9.0
Mt. McKinley	6,194	6.2
Mt. Kilimanjaro	5,895	5.9

Part C

The weight of a day pack containing hiking supplies is 20 pounds. How many ounces is this? Show your calculation.

320 oz.

Part D

Each day before climbing, it is recommended that a climber drink 1,500 milliliters of water. How many liters is this? Explain.

1.5 L

NAME _____ DATE _____

SCORE _____

Performance Task

Comparing Mountains

A scientist wants to compare information on three different mountain hikes. Research ahead of time the height in feet of Mount Everest, Mount McKinley, and Mount Kilimanjaro.

Write your answers on another piece of paper. Show all your work to receive full credit.

Part A

From your research, fill in the heights in feet of the three mountains. Then convert the measurements to miles and feet.

Mountain	Height in Feet	Height in Miles and Feet
Mt. Everest	29,029	5 miles
Mt. McKinley	20,322	3 miles
Mt. Kilimanjaro	19,341	3 miles

Performance Task (continued)

Part B

From your research, fill in the heights in meters of the three mountains. Then convert the measurements to kilometers.

Mountain	Height in Meters	Height in kilometers
Mt. Everest	8848	8
Mt. McKinley	6194	6
Mt Kilimanjaro	5895	5

Part C

The weight of a day pack containing hiking supplies is 20 pounds. How many ounces is this? Show your calculation.

$$20 \times 12 = 240$$

Part D

Each day before climbing, it is recommended that a climber drink 1,500 milliliters of water. How many liters is this? Explain.

15 liters

Student Model

Task Scenario
Students will use nets and volume and area formulas to construct a cereal box.

Part	Maximum Points	Scoring Rubric
		Depth of Knowledge DOK2, DOK3, DOK4
A	4	Full Credit: Sample Answer: 2 in. 12 in. 9 in. Partial Credit (1 point) will be given for a correct net. An additional point will be given for each of the three labels. No credit will be given for an incorrect net with incorrect labels.
B	3	Full Credit: 300 square inches; Sample answer: I multiplied 2 × 9 for one face and got 18. Then I multiplied 2 × 12 and got 24. Then I multiplied 9 × 12 and got 108. There are 2 sections with area 18, 2 sections with area 24, and 2 sections with area 108, so I added 18 + 18 + 24 + 24 + 108 + 108 and got 300. Partial Credit (1 point) will be given for a correct answer without a correct explanation. No credit will be given for an incorrect answer.

Part	Maximum Points	Scoring Rubric
C	2	Full Credit: 162 cubic inches; Sample answer: I multiplied $2 \times 9 \times 12$ and got 216 for the volume of the box. Then I multiplied $216 \times \frac{3}{4}$ and got 162 cubic inches. Partial Credit (1 point) will be given for a correct answer without a correct explanation. No credit will be given for an incorrect answer.
TOTAL	9	

Performance Task Rubrics

NAME _____ DATE _____

SCORE _____

Performance Task

Constructing Cereal Boxes

A cereal company is looking to construct a new box for their leading brand of cereal. The box and its dimensions are shown below.

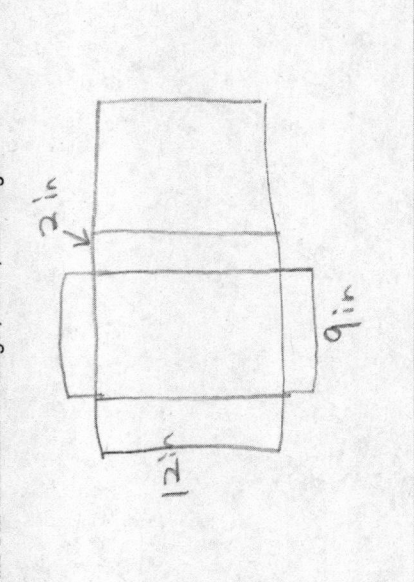

12 in

9 in

2 in

GEREAL

Write your answers on another piece of paper. Show all your work to receive full credit.

Part A

Each box needs to be cut from a flat piece of cardboard. Draw a net for the box and label the length, width, and height.

2 in

9 in

12 in

Performance Task (continued)

Part B

The amount of cardboard needed is measured in square inches. Find the area of the net you drew in Part A in order to find the area of cardboard needed to construct a box.

$2 \times 12 \times 2 = 48$

$2 \times 12 \times 9 = 216$

$2 \times 9 \times 2 = 36$

300 square inches

Part C

The box will be filled $\frac{3}{4}$ of the way with cereal. Find the volume of cereal that can be put in each box.

Volume = $2 \times 9 \times 12 = 216$

$216 \times \frac{3}{4} = 162$ cubic inches

NAME _____ DATE _____

SCORE _____

Performance Task

Constructing Cereal Boxes

A cereal company is looking to construct a new box for their leading brand of cereal. The box and its dimensions are shown below.

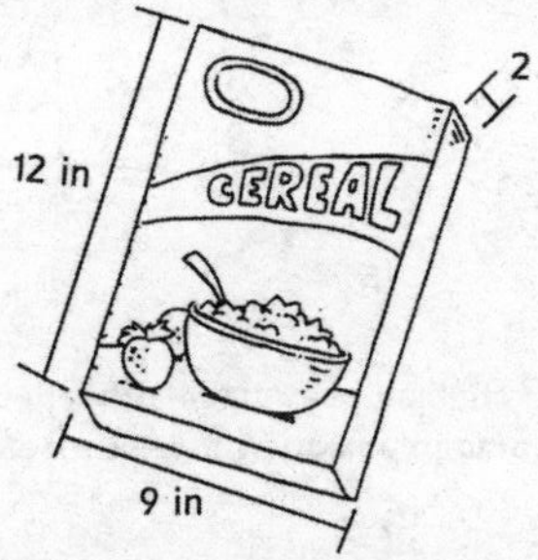

12 in

2 in

9 in

GEREAL

Write your answers on another piece of paper. Show all your work to receive full credit.

Part A

Each box needs to be cut from a flat piece of cardboard. Draw a net for the box and label the length, width, and height.

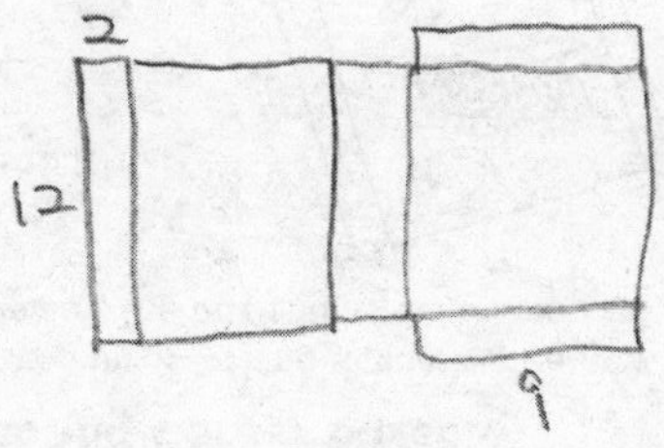

2

12

Performance Task (continued)

Part B

The amount of cardboard needed is measured in square inches. Find the area of the net you drew in **Part A** in order to find the area of cardboard needed to construct a box.

$2 \times 12 = 24$

$2 \times 9 = 18$

$9 \times 12 = 108$

$$\begin{array}{r} 148 \\ \times\ 2 \\ \hline 296 \text{ square inches} \end{array}$$

Part C

The box will be filled $\frac{3}{4}$ of the way with cereal. Find the volume of cereal that can be put in each box.

$2 \times 9 \times 12 \times \frac{3}{4}$

162

Student Model

NAME _____ DATE _____

SCORE _____

Performance Task

Constructing Cereal Boxes

A cereal company is looking to construct a new box for their leading brand of cereal. The box and its dimensions are shown below.

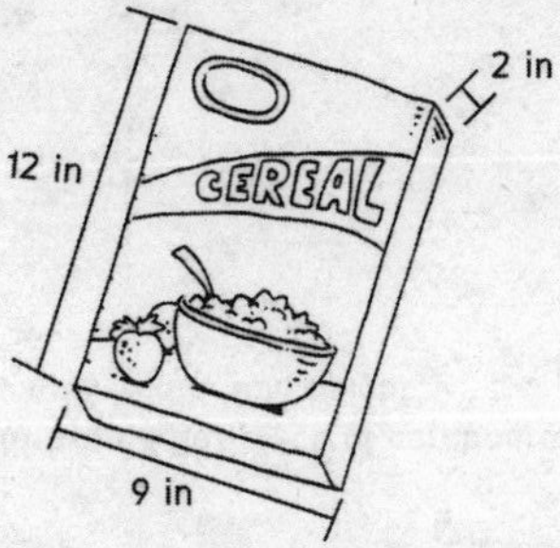

12 in

2 in

9 in

CEREAL

Write your answers on another piece of paper. Show all your work to receive full credit.

Part A

Each box needs to be cut from a flat piece of cardboard. Draw a net for the box and label the length, width, and height.

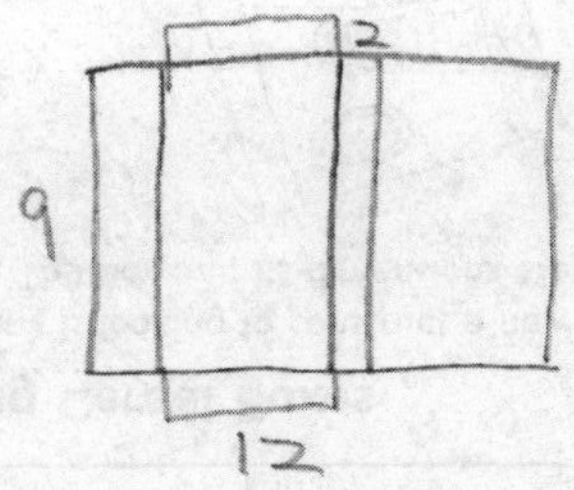

2

12

9

Performance Task (continued)

Part B

The amount of cardboard needed is measured in square inches. Find the area of the net you drew in Part A in order to find the area of cardboard needed to construct a box.

$2 \times 9 \times 12 \times 2 = 432$ square inches

Part C

The box will be filled $\frac{3}{4}$ of the way with cereal. Find the volume of cereal that can be put in each box.

$2 \times 9 \times 12 = 216$ cubic inches

NAME _____ DATE _____

SCORE _____

Performance Task

Constructing Cereal Boxes

A cereal company is looking to construct a new box for their leading brand of cereal. The box and its dimensions are shown below.

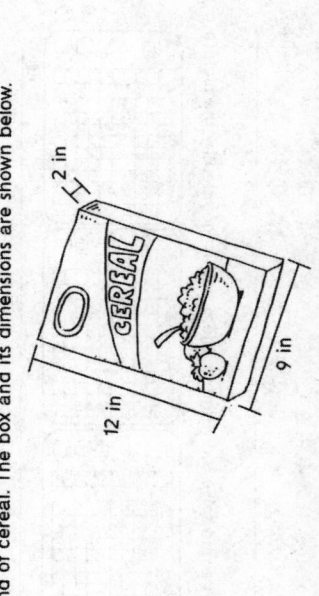

Write your answers on another piece of paper. Show all your work to receive full credit.

Part A

Each box needs to be cut from a flat piece of cardboard. Draw a net for the box and label the length, width, and height.

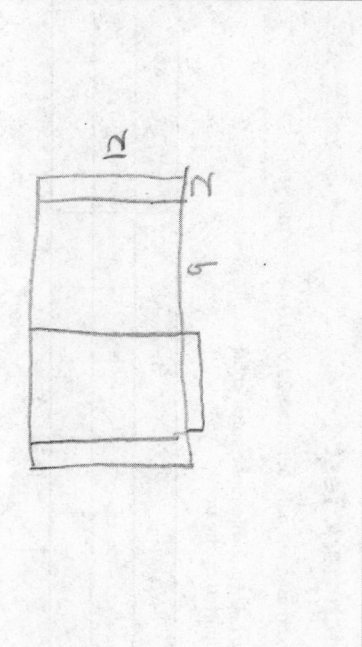

Performance Task *(continued)*

Part B

The amount of cardboard needed is measured in square inches. Find the area of the net you drew in **Part A** in order to find the area of cardboard needed to construct a box.

$$2 \times 12 = 24$$
$$2 \times 9 = 18$$
$$9 \times 12 = 108$$
$$2 \times 12 = 24$$
$$9 \times 12 = 108$$
$$\overline{282}$$

Part C

The box will be filled $\frac{3}{4}$ of the way with cereal. Find the volume of cereal that can be put in each box.

volume = length x width x height

216

Student Model

NAME

DATE

Benchmark Test 1

SCORE

1. An amusement park admitted 28,512,121 people last year.

Part A: Fill in the place value chart for the number of people admitted by the park.

Millions			Thousands			Ones		
hundreds	tens	ones	hundreds	tens	ones	hundreds	tens	ones
	2	8	5	1	2	1	2	1

20,000,000   8,000,000   500,000   10,000   2,000   100   20   1

Part B: Write the number in words.

Twenty-eight million, five hundred twelve thousand, one hundred twenty-one.

Part C: Write the expanded form of the number.

$2 \times 10,000,000 + 8 \times 1,000,000 + 5 \times 100,000 + 1 \times 10,000 + 2 \times 1,000 + 1 \times 100 + 2 \times 10 + 1$

2. Write and solve a division problem that is modeled by the picture.

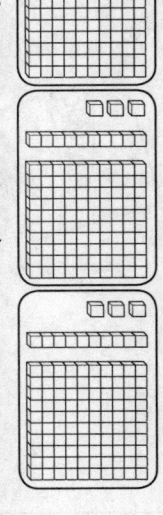

$340 \div 3 = 113 \text{ R } 1$

3. The Schmidt family is decorating their house for the winter. Packages of lights cost the same amount, but they have different size strands in them. Mr. Schmidt want to buy the package that has the greatest number of lights.

Light Bulb Packages	
Strands in Package	Light Bulbs Per Strand
5	800
10	350
15	260
20	210
25	160

Part A: Complete the table below with the number of lights per package.

Strands in a Package	5	10	15	20	25
Lights Per Package	4,000	3,500	3,900	4,200	4,000

Part B: Which package provides the greatest number of light bulbs?

The package with 20 strands.

4. A surveyor is dividing a large plot of land that measures 27,512 square miles. He wants to divide the area into 91 equal regions. Estimate how large each area will be. Show your work.

$27,000 \div 90 = 300$ square miles

5. Which of the following three is not equal to the other two? Circle the answer.

5.62

(Five and sixty-two tenths)

$5 \times 1 + 6 \times \frac{1}{10} + 2 \times \frac{1}{100}$

Left page (151)

6. The chart shows the cost of several school supplies. Which combinations can you buy with $23?

Pencil	$1
Notepad	$5
Binder	$7
Pen	$2

Yes	No	
☐	☐	2 binders, 1 notepad, and 2 pens
☐	☐	3 binders and 2 pencils
☐	☐	3 notepads, 1 binder, 1 pen, and 1 pencil
☐	☐	1 notepad, 1 binder, 3 pens, and 5 pencils

7. Nine families went on a campout together. The total bill for the weekend supplies was $603. The families will split the bill evenly.

Part A: How much should each family contribute?

$67

Part B: Estimate to check your answer.

$600 ÷ 10 = $60. The answer is reasonable.

8. A local car dealership sells 9,792 cars per year. How many cars does the dealership sell per month?

816 cars

Right page (152)

9. The following chart lists the weight of six packages that came into the post office. Place the weights in order from least to greatest.

3.15 lbs	3.51 lbs	3.05 lbs
5.03 lbs	5.13 lbs	3.015 lbs

3.015, 3.05, 3.15, 3.51, 5.03, 5.13

10. The product of 76 and another number is 15,580. Complete the table to help you estimate the other number.

76 × 100	7,600
76 × 150	11,400
76 × 200	15,200
76 × 250	19,000
76 × 300	22,800

The other number is about 200.

11. *Part A*: Fill in the following table with quotients and remainders.

Division Problem	Quotient	Remainder
5,338 ÷ 13	410	8
5,337 ÷ 13	410	7
5,336 ÷ 13	410	6

Part B: What pattern do you notice?

As the dividend goes down by 1, the remainder goes down by 1.

12. Jayna drives 17 miles each day to work. She wants to know how many miles she drives in a month. Is there too much information or not enough information to solve this problem? Shade the box next to the correct description. If there is too much information, name the extra information and solve the problem. If there is not enough information, describe what Jayna would need to know to solve the problem.

☐ Too much information   ▣ Not enough information

Jayna would need to know how many days per month she drives.

13. Jenna is trying to use the digits 1, 2, 2, 0, 2, 1 to make a number that is between 210,000 and 220,000. Shade the box next to any answer that is correct.

☐ 122,021   ☐ 210,221

▣ 210,121   ☐ 222,110

14. The Chen family is saving for a vacation to Europe. They need $7,000 for the trip. The family plans to save $312 per month.

Part A: Fill in the partial product diagram to show how much they will save in two years.

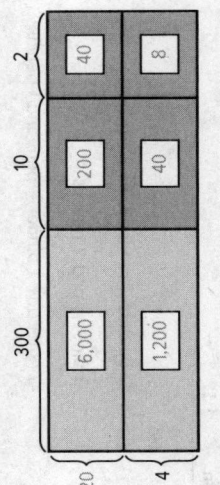

	300	10	2
20	6,000	200	40
4	1,200	40	8

$7,488

Part B: How much extra money will they have saved?

$488

15. An apple farmer sells apples in bag that hold 7 apples. Her picking crew has picked all the apples that are ready for the weekend sale and begin packaging them into the bags of 7. At the end of the bagging, they have some leftover apples. Shade the boxes next to any number that is a possible remainder, then explain your reasoning.

☐ 0   ▣ 3   ☐ 6
☐ 1   ▣ 4   ☐ 7
☐ 2   ▣ 5   ☐ 8

When dividing by 7, the only remainders possible are numbers that are less than 7.

16. A local charity is storing up large containers of drinking water for emergency purposes. Each container of water costs $19. The charity has $5,147 in donations.

Part A: How many containers of water can the charity buy?

270 containers

Part B: What is the remainder, and what does it mean?

$17. This is the amount of money they have left over.

Part C: Estimate to check your answer. Show your work.

$5,100 ÷ $20 = 255 containers. The answer is reasonable.

17. The table shows the number of pounds of sugar that a bakery used in three different months. If the bakery used 28 fewer pounds in January than they did in February and the total pounds for the three months was 726 miles, fill in the missing values on the chart.

Month	Pounds of Sugar
January	223
February	251
March	252

18. The following table shows the number of toothpicks in several different boxes together with the number of boxes in a package. Fill in the table with the missing values.

Number of Toothpicks in a Box	Number of Boxes in a Package	Total Number of Toothpicks in a Package
10^2	85	8,500
10^2	125	12,500
10^3	176	176,000
10^4	298	2,980,000

19. A pizza company is open 50 weeks per year. In one year, they sold 8,350 pizzas.

Part A: Fill in the division fact with compatible numbers to estimate the average number of pizzas the company sold per week.

$8,000$ ÷ 50 = 160 pizzas

Part B: Find the exact average of pizzas sold per week.

167 pizzas

Part C: Is your estimate greater than or less than the actual number? Explain how you could have known this ahead of time.

Sample answer: The estimate is less than the actual number. I could have known this would be true because I rounded 8,350 down to 8,000 for the estimate.

20. Circle any that would not be good ways of estimating 4,924 ÷ 71.

4,900 ÷ 70 = 70   5,000 ÷ 100 = 50

4,000 ÷ 100 = 40   4,000 ÷ 50 = 80

Task Scenario
Students will use multiplication, division, and rounding to plan the number of days to drive across the country.

Depth of Knowledge	DOK2, DOK3

Part	Maximum Points	Scoring Rubric
A	3	**Full Credit:** | Distance from New York to Chicago	800 miles
Distance from Chicago to LA	2,000 miles
Total	2,800 miles	 Partial Credit (1 point each) will be given for the correct distances with an incorrect total. No credit will be given for having an incorrect total and incorrect distances.
B	2	**Full Credit:** 520 miles per day; Sample answer: $8 \times 65 = 520$ Partial Credit (1 point) will be given for a correct answer without an explanation. No credit will be given for an incorrect answer.
C	3	**Full Credit:** 9 days; Sample answer: To Chicago, they will spend 2 days because $800 \div 520 = 1\text{ R }280$. To Los Angeles, they will spend 4 days because $2,000 \div 520 = 3\text{ R }440$. The total trip is $2 + 4 + 3 = 9$ days. Partial Credit (1 point each) will be given for having the correct number of days for each leg of the trip, but an incorrect total. No credit will be given for incorrect durations for the parts of the trip and an incorrect total.
TOTAL	8	

NAME _____ DATE _____

Benchmark Test 2

SCORE _____

1. Round the following number to the nearest hundredth. Write the rounded number in both expanded form and standard form.

$$4 \times 10 + 3 \times 1 + 7 \times \frac{1}{10} + 8 \times \frac{1}{100} + 6 \times \frac{1}{1000}$$

Expanded Form

$$4 \times 10 + 3 \times 1 + 7 \times \frac{1}{10} + 9 \times \frac{1}{100}$$

Standard Form

43.79

2. A cook uses 0.4 pounds of butter every morning on croissants. Regroup and shade the models to figure out how much butter he uses in five days.

Monday   Tuesday   Wednesday   Thursday   Friday

2.0 _____ pounds

---

3. A rock is thrown up in the air. The height of the rock in feet after three seconds is $4 \times 3^2 + 6 \times 3 + 12$. Find the height of the rock.

66 feet

4. Jamal ran 21 miles in 5 days. Circle any of the following that describe the average number of miles Jamal ran per day.

$\frac{5}{21}$ $\frac{21}{5}$ $4\frac{1}{5}$ $5\frac{1}{4}$

5. A discount book club has a monthly fee of $6. Once you pay the fee, you can buy books for $5 each. In January, Mr. Huan joined the book club and bought 12 books.

Part A: Write an expression for how much Mr. Huan spent.

$6 + 5 \times 12$

Part B: Evaluate the expression to find out how much John spent.

$66

6. Put the following numbers in order from least to greatest.

54.03×10^3 5.403×10^2 0.0543×10^4 54.03×10^3

5.403×10^2 0.0543×10^4 54.03×10^3

Benchmark Tests

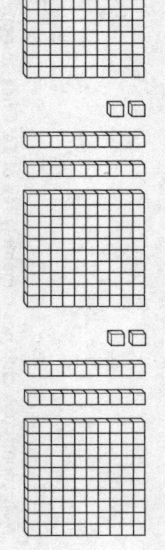

7. Consider the following five numbers.

10   ~~15~~   20   ~~25~~   30

Part A: What is the greatest common factor of all five numbers?

5

Part B: Cross out two of the five numbers so that the remaining numbers have a greatest common factor of 10.

8. Write the correct property of addition for each step.

4.4 + (3.2 + 2.6) + 0

= 4.4 + (2.6 + 3.2) + 0   **Commutative Property**

= (4.4 + 2.6) + 3.2 + 0   **Associative Property**

= 7.0 + 3.2 + 0   Addition

= 10.2 + 0   Addition

= 10.2   **Identity Property**

9. A factory needs to sell $\frac{2}{3}$ of its inventory in order to make a profit. The fractions below represent the part of the factory's inventory that was sold. Circle any fractions that represent the factory making a profit.

$\frac{7}{11}$ (⃝$\frac{9}{13}$) $\frac{10}{16}$

10. An electrician has purchased 102 meters of wire. Each time he wires an outlet in a particular room he uses 15.17 meters of wire. Fill in the following table to figure out how many outlets he can wire and how much wire will be left over.

Outlets Wired	Wire left over
1	102 − 15.17 = 86.83 meters
2	71.66 meters
3	56.49 meters
4	41.32 meters
5	26.15 meters
6	10.98 meters
7	

__6__ outlets

__10.98__ meters left over

11. Giovanni was given the following diagram that is supposed to represent a decimal division problem. Write the problem, and find the answer.

3.66 ÷ 3 = 1.22

12. Circle the pattern that does not belong.

3, 6, 9, 12

2, 5, 8, 11

(2, 4, 8, 16)

2, 6, 10, 14

15. Shade the box under "Yes" or "No" to indicate whether each problem will require regrouping.

Yes	No	
☐	■	7.21 + 0.73
☐	■	29.13 + 20.35
■	☐	35.05 + 27.05
■	☐	121.92 + 2.18

16. Glow sticks are sold in packs of various sizes. Which of the three brands is the best buy?

Brand	Number in Pack	Price
A	6	$3.57
B	15	$8.25
C	20	$11.20

Brand B

17. The table below shows the number of marbles of three different colors that a marble shop has. They want to package the marbles into bags that will have only one color of marble, and they want to make sure that each bag has the same number of marbles. If all of the marbles are put into bags, what is the greatest number of marbles that could be in each bag?

Pink	75
Red	30
White	60

15 marbles

---

13. Match each fraction with its decimal equivalent.

$\frac{11}{25}$ 0.60

$\frac{3}{5}$ 0.75

$\frac{3}{20}$ 0.44

$\frac{3}{4}$ 0.15

14. A bicyclist is riding around town for the morning. The map below shows the places he visits.

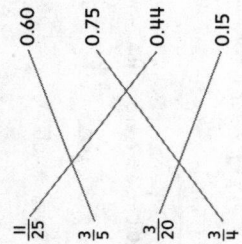

Part A: If the bicyclist starts at home, describe the path he can take to the bookstore, and then to the coffee shop.

6 units south, then 5 units west to the bookstore. 5 units west, then 3 units north to the coffee shop.

Part B: The bicyclist goes directly home from the coffee shop. How many units was his total trip?

32 units

Benchmark Tests

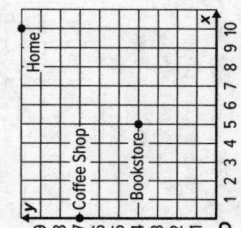

18. A local fundraising effort managed to raise the following dollar figures in three different activities.

Car Washes	$598
Bake Sales	$243
Yardwork	$102

Part A: Write down the best order in which to add the numbers so that it is easiest to find the total using mental math.

$598 + $102 + $243

Part B: Find the total amount raised.

$943

19. A soccer coach has marked off an area in the shape of a triangle for some soccer drills. He wants to put a cone at every point that lies *inside* the triangle. Circle any coordinate pair on which the coach should put a cone.

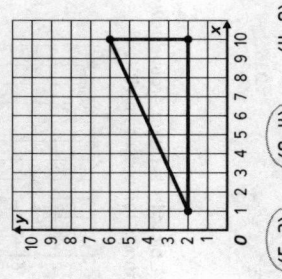

(5, 3)   (8, 4)   (4, 8)   (9, 4)   (3, 7)

20. Circle the expression that is not equal to the others.

988 ÷ 1,000   9.88 ÷ 10

98.8 ÷ 100   98.8 ÷ 10,000

Page 164 • Apple Picking

Task Scenario
Students will use multiplication, division, addition, and subtraction to calculate the cost of apples and the number of pies that can be made.

Depth of Knowledge	DOK2, DOK3	
Part	**Maximum Points**	**Scoring Rubric**
A	2	Full Credit: There are 2 gallons in a peck, so there are 10.5 gallons of apples. Partial Credit (1 point) will be given for the number of gallons in a peck. No credit will be given for an incorrect number of gallons.
B	2	Full Credit: 1.25 pecks; Sample answer: I subtracted 5.25 − 1.5 and got 3.75. Since they have 2.5 pecks left, I subtracted 3.75 − 2.5 and got 1.25 pecks. Partial Credit (1 point) will be given for a correct answer without a reasonable explanation. No credit will be given for an incorrect answer.
C	2	Full Credit: $28.35; Sample answer: I multiplied 5.40 by 5.25 and got 28.35. Partial Credit (1 point) will be given for a correct answer without a reasonable explanation. No credit will be given for an incorrect answer.
D	2	Full Credit: 10 pies; Sample answer: I divided 2.5 by 0.25 and got 10. Partial Credit (1 point) will be given for a correct answer without a reasonable explanation. No credit will be given for an incorrect answer.
TOTAL	8	

Performance Task Rubrics

NAME _____ DATE _____

Benchmark Test 3   SCORE _____

1. Juanita has three more than 2 times the number of books than her friend Uma has.

Part A: If Uma has 6 books, write an expression for the number of books Juanita has. Then find the number of books.

$2 \times 6 + 3 = 15$ books

Part B: If Uma has 7 books, how would the number of books in Juanita's collection compare to your answer in **Part A?** Explain.

Juanita would have 2 more books, or 17 since the number multiplied by the 2 would increase by 1.

2. Mr. Ortiz is distributing tennis supplies to his team of 20 players. The extras he will store for future years. Write an expression for the number of supplies each player receives and then evaluate each expression.

Supplies	Expression	Each Player Receives
95 tennis balls	$95 \div 20$	4
61 racket grips	$61 \div 20$	3
1,200 bottles of water	$1,200 \div 20$	60
24 t-shirts	$24 \div 20$	1

3. Which of the following can be modeled by the division expression $550 \div 5$? Choose all that apply.

A. 550 dollars distributed evenly to 5 groups

B. 5 points distributed evenly 550 times

C. 550 pounds distributed evenly into bags of 5 pounds each

D. 550 feet per step for 5 steps

---

4. Compare $\frac{7}{10}$ and $\frac{7}{100}$.

Part A: Shade the decimal models to show each fraction. Then write each as a decimal.

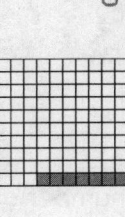

0.7

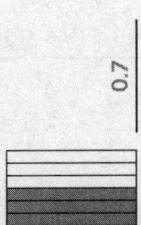

0.07

Part B: Compare the two decimals. Use >, <, or =. Explain.

$0.7 > 0.07$; 7 parts out of 10 is greater than 7 parts out of 100.

5. A farmer measured a sample of his corn plants after two weeks and recorded the following measurements in inches: $4\frac{1}{2}$, 5, $6\frac{3}{4}$, $7\frac{1}{4}$, $4\frac{1}{2}$, $6\frac{1}{4}$, 7, $6\frac{3}{4}$, 7, $6\frac{3}{4}$.

Part A: Use the line plot to record the measurements.

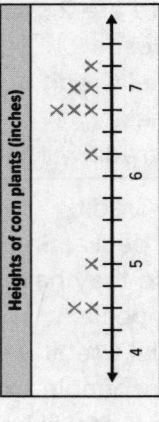

Heights of corn plants (inches)

Part B: What is the sum of the heights of all the corn plants?

$61\frac{3}{4}$ inches

6. Each dog in a kennel needs 8 pounds of food for their upcoming stay. If the kennel has 225 pounds of food, will there be enough to accommodate 20 dogs? If so, how many more dogs can be accommodated? Explain your reasoning.

Yes; 8; Sample answer: $20 \times 8 = 160$ and $160 < 225$; $225 \div 8 = 28$ R1; $28 - 20 = 8$

7. Draw the decimal points on each number on the left side of the equation so that the difference is correct as shown.

$$3\,2\,5 - 4\,2 = 28.3$$

8. Shanna is delivering papers on her morning route. She starts at (0, 0), and three of her houses are at points 3 right, 6 up; 2 up, 1 right; and 5 right, 3 up. Use the grid to draw and label the points where the houses are located.

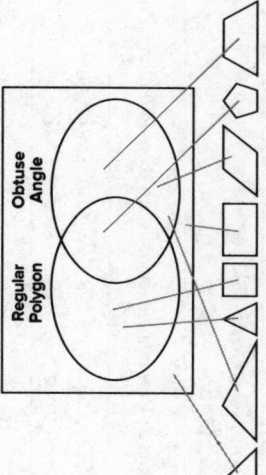

9. While dividing numbers with zeros at the end, Ethan notices a certain pattern. His results are shown in the table.

Expression	Quotient
100 ÷ 5	20
1000 ÷ 50	20
10 ÷ 5	2
100 ÷ 50	2

Part A: What pattern does he recognize?

Sample answer: The first digit is always 2, and the number of zeros is equal to how many more zeros the dividend has than the divisor, minus 1.

Part B: Using this pattern, what is the result of 10,000 ÷ 50?

200

10. In a competition for the tallest stack of balanced blocks, the top four heights were recorded as follows: 2.9 m, 2.77 m, 2.81 m, 2.84 m. Place a dot on the number line for each given height and label.

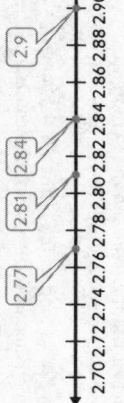

2.77   2.81   2.84   2.9

2.70 2.72 2.74 2.76 2.78 2.80 2.82 2.84 2.86 2.88 2.90

---

11. A carpenter is cutting a piece of wood that is 1 yard by 1 yard. He wants one side to be $\frac{3}{8}$ of a yard and one side to be $\frac{1}{4}$ of a yard.

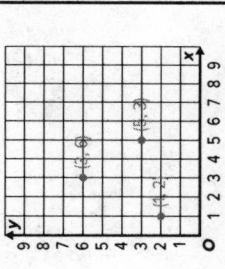
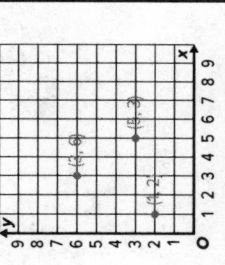

Part A: Model the desired cut on the square of wood shown.

Part B: What is the area of the wood?

$$\frac{3}{8} \times \frac{1}{4} = \frac{3}{32} \text{ square yards}$$

12. Use the Venn diagram to sort the shapes. Draw a line from the shape to the correct area of the diagram.

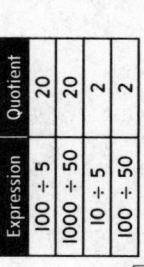

Regular Polygon   Obtuse Angle

13. A group of friends earned $5,520 doing yard work around town for a year. They decided to give one third of the money to a charity. Then each friend received an equal portion of the money, but less than what was given to charity. Were there 2, 3, or 4 friends? How much money did each friend receive? Explain your reasoning.

4 friends; $920; Sample answer: $\frac{1}{3} \times 5{,}520 = 1{,}840$; $5{,}520 - 1{,}840 = 3{,}680$; $3{,}680 \div 2 = 1{,}840$, which is not less than the charity amount. $3{,}680 \div 3$ cannot be divided evenly. $3{,}680 \div 4 = 920$. This is the answer.

14. Julian has 12 unit blocks. He needs to create rectangular prisms.

Part A: Use the table shown to create as many rectangular prisms as you can.

Part B: As long as the prism has the same three numbers for the sides, it is considered to be the same. How many unique prisms can you create?

Length	Width	Height
1	1	12
1	2	6
1	3	4
2	2	3

4

15. Draw lines between equivalent numbers.

$1\frac{6}{7}$ $3\frac{2}{5}$

$\frac{2}{3}$ $\frac{16}{24}$

$\frac{17}{5}$ $\frac{18}{24}$

$\frac{13}{7}$

16. The order for a collection of artwork is shown. What is the total area of canvas needed to for the order of all of the artwork?

Artwork Order		
Quantity	Dimensions (in.)	Total Square Inches
6	$12\frac{3}{4} \times 8$	612
2	$10\frac{3}{8} \times 10$	$207\frac{1}{2}$
8	$15\frac{1}{2} \times 9$	1,116

Total = $1935\frac{1}{2}$ square inches

17. Which of the following use the proper order of operations? Select all that apply.

- ☑ $20 - (6 + 5 \times 2) = 38$
- ☐ $4 \times [8 - (10 \div 2)] = 12$
- ☐ $15 - 1 + 3 = 11$
- ☑ $4^2 + 2 \times 5 = 26$
- ☐ $2 \times (24 + 1) = 50$

18. There are 10 milligrams in every centigram, 100 milligrams in every decigram, and 1000 milligrams in every gram.

Part A: How many centigrams are there in a gram? Explain.

Sample answer: 100 centigrams in a gram; One centigram is $\frac{1}{100}$ of a gram.

Part B: How many decigrams are there in a gram? Explain.

Sample answer: 10 decigrams in a gram; One decigram is $\frac{1}{10}$ of a gram.

Part C: How does a gram compare to a decigram? Explain.

A gram is 10 times greater than a decigram; 10 decigrams make a gram.

19. While trying to add lengths for marking off a garden plot, Mrs. Shen calculated that $\frac{3}{5} + \frac{1}{2} = \frac{4}{7}$. Use estimation to determine if Mrs. Shen is correct.

Sample answer: $\frac{3}{5}$ is more than $\frac{1}{2}$. Since $\frac{1}{2} + \frac{1}{2} = 1$, you would expect the answer to be over 1. She is not correct.

20. Martina compared numbers with similar digits. Using mathematical language, explain how each set of numbers is different.

A	12,678 and 2,678
B	576 and 57.6
C	49 and 049

Sample answer:
A: The second number is 10,000 less than the first number.

B: The first number is a whole number. It represents 576 ones. The second number is a decimal fraction. It represents 57 ones and 6 tenths of 1 whole.

C: Both numbers have the same value. When a zero is placed as the first digit in a whole number, it has no value.

21. At the end of a night, a cashier empties the registers of the $1, $10, and $100 bills. There are 40 bills total. Which of the following would make the least amount of money? Explain how you solved the problem.

Possible Bill Combinations		
$1 Bills	$10 Bills	$100 Bills
6	34	0
4	35	1
16	22	2
3	36	1

If the cashier had 6 ones, 34 tens, and 0 hundreds, she would have the smallest amount of $346. I multiplied each column by the appropriate power of 10 and then added across the rows.

22. Pianos have 88 keys. If a company produces 620 keys, how many pianos can be produced? What does the remainder mean in this case?

7 pianos; 620 divided by 88 is 7 R4. The remainder of 4 represents the number of keys left over after the 7 pianos have been made.

23. Belinda's work is shown for a recent test on fraction operations.

$$7 \div \frac{1}{5} = \frac{7 \div 1}{5} = \frac{7}{5}$$

Part A: What mistake did Belinda make while dividing?

She divided the 7 by the 1 instead of dividing it by $\frac{1}{5}$.

Part B: What is the correct quotient?

35

24. A long jumper jumped 19.569 feet.

Part A: Round this number to the nearest hundredth.

19.57 feet

Part B: Place rounded number on the number line shown.

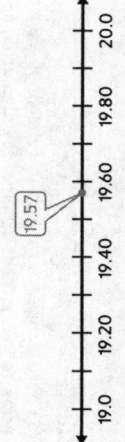

19.57

19.0 19.20 19.40 19.60 19.80 20.0

25. Explain how you model 0.6 × 0.5 using a decimal grid. How would this vary from 1.6 × 0.5?

Sample answer: For 0.6 × 0.5, I would use a 10 by 10 grid to show 6 rows by 5 columns; For 1.6 × 0.5 I would show 10 rows by 5 columns and another 6 rows by 5 columns.

26. While trying to solve 4.21 + 0.52, Raj found the sum to be 42.62. Explain why this is or is not a valid answer using estimation.

Sample answer: This is not a valid answer because 4.21 is less than 5 and 0.52 is less than 1, so the answer cannot be more than 6.

27. A baker gradually added flour to a mixing bowl for bread. He started with 3 cups of flour. He then added 1.5 cups. The third time he added 5 ounces less than the second time. Finally he added 5 ounces. How many cups of flour did he add in all? Explain your reasoning.

6 cups; Sample answer: 3 cups is the same as 24 ounces. 1.5 cups is the same as 12 ounces. The third time is 5 fewer ounces, so 7 ounces. The total is 24 + 12 + 7 + 5 = 48 ounces. 48 ounces is 6 cups.

28. A piece of wire is 2.4 meters in length. An electrician can create 4 equal sized wires from this piece or 6 equal sized wires from this piece. Use the number lines below to model each option.

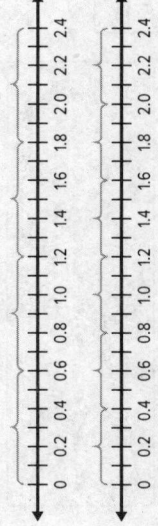

0 0.2 0.4 0.6 0.8 1.0 1.2 1.4 1.6 1.8 2.0 2.2 2.4

0 0.2 0.4 0.6 0.8 1.0 1.2 1.4 1.6 1.8 2.0 2.2 2.4

29. Juan went on four runs over the past week. The first two measured $\frac{3}{4}$ mile and $1\frac{1}{4}$ miles. The third and fourth measured $\frac{7}{8}$ mile and $1\frac{3}{8}$ miles.

Part A: Find the sum of the first two, then the sum of the last two.

First two: 2 miles; Last two: $2\frac{5}{8}$ miles

Part B: Describe how the adding processes differed.

The first set are like fractions, so I just needed to add the numerators. The second set I needed to find equivalent fractions before adding.

Part C: What is the total of all four runs?

$4\frac{5}{8}$ miles

30. A restaurant owner is buying packages of toothpicks. Each package comes with a certain number of boxes that contain the toothpicks. Because they all cost roughly the same price, he decides he wants the package that provides the most toothpicks.

Part A: Complete the table

Toothpick Packages		
Number of Boxes in a Package	Toothpicks per Box	Total Toothpicks
2	800	1,600
3	450	1,350
5	350	1,750
10	160	1,600

Part B: Which package provides the greatest number of toothpicks?

the package with 5 boxes

31. You have a recipe for a fruit smoothie, but you want to increase the recipe so that it feeds $3\frac{1}{2}$ times the original number of people. Complete the table by writing the new measurements.

Original Recipe	$3\frac{1}{2}$ Times Recipe
1 cup bananas	$3\frac{1}{2}$ cups bananas
$2\frac{1}{2}$ cups strawberries	$8\frac{3}{4}$ cups strawberries
$\frac{3}{4}$ cup of milk	$2\frac{5}{8}$ cups of milk
2 tablespoons of honey	7 tablespoons of honey

Page 175 • Building a Corn Bin

Task Scenario Students will create a line plot and will use the formula for the volume of a rectangular prism applied to prisms with fractional side lengths to model a corn bin.		
Depth of Knowledge	DOK2, DOK3	

Part	Maximum Points	Scoring Rubric
A	2	Full Credit: **Lengths of Beams (feet)** No credit will be given for an incorrect answer.
B	2	Full Credit: 384 inches; Sample answer: $4 \times (8 - 6\frac{1}{2}) + 4 \times (15 - 12\frac{1}{2}) + 4 \times (20 - 16) = 6 + 10 + 16 = 32$ feet. Multiply the number of feet by 12 to find the number of inches. 32 feet is 384 inches. Partial Credit (1 point) will be given for a correct answer in feet but an incorrect or missing conversion to inches. No credit will be given for an incorrect answer.
C	2	Full Credit: 1,300 cubic feet; $12\frac{1}{2} \times 16 \times 6\frac{1}{6} = 1,300$ Partial Credit (1 point) will be given for a correct answer without an appropriate calculation. No credit will be given for an incorrect answer.
D	2	Full Credit: The volume of the new bin will be twice the volume of the original bin. The volume of the original bin is $12\frac{1}{2} \times \ell \times h$. The volume of the new bin will be $25 \times \ell \times h$, so the volume of the new bin is twice the volume of the original bin. Partial Credit (1 point) will be given for stating the correct relationship between the volumes without a correct explanation. No credit will be given for an incorrect answer.
TOTAL	8	

Performance Task Rubrics

Benchmark Test 4

NAME _____ DATE _____ PERIOD _____

SCORE _____

Benchmark Test 4

1. Mr. Li is buying packages of notebooks. Because they all cost roughly the same price, he decides he wants the package that provides the most sheets of paper.

Notebook Package Sizes

Number of Notebooks	Sheets per package
1	800
3	350
6	250
10	150

Part A: Complete the table below with the number of total sheets per package.

Notebooks	Total Sheets
1	800
3	1050
6	1500
10	1500

Part B: Which package provides the most total sheets of paper? Justify your response.

6-notebook or 10-notebook package; both of these packages provide 1500 sheets of paper.

2. Compare each number to 85.2. Use the symbols <, >, or =.

85.20 $=$ 85.2

852 + 0.2 $>$ 85.2

eighty-five ones and two tenths $=$ 85.2

852 hundredths $<$ 85.2

3. In a competition for the tallest sunflower, the top four heights were recorded as follows: 3.9 m, 3.88 m, 3.82 m, 3.76 m.

Place a dot on the number line and label for each given distance.

3.70 3.72 3.74 3.76 3.78 3.8 3.82 3.84 3.86 3.88 3.90

4. Julian compared numbers with similar digits. Using mathematical language, explain how each set of numbers is different.

A	13.542 and 35.42
B	781 and 78.1
C	1.2 and 01.20

Sample answer:

A: The second number is greater than the first number because there are 35 wholes and part of another. The first number has only 13 wholes and part of another.

B: The first number is a whole number. It represents 781 ones. The second number is a decimal fraction. It represents 78 ones and 1 tenth of 1 whole.

C: Both numbers have the same value. When a zero is placed as the first digit in a whole number, it has no value. When a zero is placed as the last digit in a decimal, it has no value.

Benchmark Test 4

5. There are 10 millimeters in every centimeter, 100 millimeters in every decimeter, and 1000 millimeters in every meter.

Part A: How many centimeters are there in a meter? Explain.

There are 100 centimeters in a meter; One centimeter is $\frac{1}{100}$ of a meter.

Part B: How many decimeters are there in a meter? Explain.

There are 10 decimeters in a meter; One decimeter is $\frac{1}{10}$ of a meter.

Part C: How does a meter compare to a decimeter? Explain.

A meter is 10 times greater than a decimeter; 10 decimeters make a meter.

6. Compare $\frac{5}{10}$ and $\frac{5}{100}$.

Part A: Shade the decimal models to show each fraction. Then write each as a decimal.

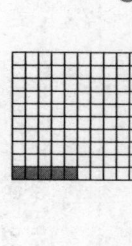

0.5

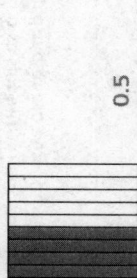

0.05

Part B: Compare the two decimal fractions. Use >, <, or =. Explain.

$\frac{5}{10}$ (>) $\frac{5}{100}$; 5 parts out of 10 is greater than 5 parts out of 100.

7. While trying to add together the widths of some boards for a deck he is building, Haj said that $\frac{4}{7} + \frac{1}{2} = \frac{5}{9}$. Use estimation to determine if Haj is correct.

Sample answer: $\frac{4}{7}$ is more than $\frac{1}{2}$. Since $\frac{1}{2} + \frac{1}{2} = 1$, you would expect the answer to be over 1. Since $\frac{5}{9} < 1$, Haj is incorrect.

Grade 5 • Benchmark Test 4   179

8. An academic team won a scholarship of $5,856. One fourth of the money was first given to a charity. Then each team member received an equal portion of the scholarship money, but less than what was given to charity. Were there 2, 3, or 4 team members? How much money did each team member receive? Explain your reasoning.

There were 4 team members; $1,098 each; Sample answer: $5,856 × $\frac{1}{4}$ = $1,464 given to charity; $5,856 − $1,464 = $4,392 left to divide; $4,392 ÷ 4 = $1,098 for each team member.

9. Which of the following can be modeled by the division expression 650 ÷ 5? Choose all that apply.

A. $650 distributed evenly to 5 groups

B. 5 points distributed evenly 650 times

C. 650 ounces distributed evenly into beakers of 5 ounces each

D. 650 millimeters per step for 5 steps

10. Each student needs 11 pencils for the school year. If the school started with a box of 1,325, would there be enough for a school of 120? If so, how many more students can be supplied pencils? Explain your reasoning.

Yes; no more students can be supplied; Sample answer: 120 × 11 = 1,320; 1,325 ÷ 11 = 120 R5.

11. Ms. Marcella is distributing school supplies to her classroom of 30 students. Write an expression for the number of supplies each student receives and then evaluate each expression.

Supplies	Expression	Each Student receives
100 folders	100 ÷ 30	3
65 highlighters	65 ÷ 30	2
1,000 sheets of paper	1,000 ÷ 30	33
34 calculators	34 ÷ 30	1

180   Grade 5 • Benchmark Test 4

Benchmark Tests

Grade 5 • Benchmark Test 4

337

12. While dividing two numbers with zeros at the end, Eugenio notices a certain pattern. His results are shown in the table.

Expression	Quotient
100 ÷ 2	50
1000 ÷ 20	50
10 ÷ 2	5
100 ÷ 20	5

Part A: What pattern does he recognize?

Sample answer: The first digit is always 5, and the number of zeros is equal to how many extra zeros the dividend has minus 1.

Part B: Using this pattern, what is the result of 10,000 ÷ 20?

500

13. If 365 work days are to be split among 12 employees evenly, how can you rewrite this as a fraction? What does the fraction represent?

$\frac{365}{12}$. Sample answer: The fraction represents the quotient or the number of days each worker will work.

14. A grasshopper jumped 6.434 centimeters.

Part A: Round this number to the nearest hundredth.

6.43

Part B: Place the rounded number on the number line shown.

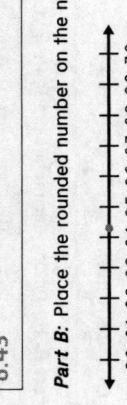

6.0 6.1 6.2 6.3 6.4 6.5 6.6 6.7 6.8 6.9 7.0

15. While trying to add 1.43 + 0.71, Kevin found the sum to be 15.01. Use estimation to explain why this is or is not a valid answer.

This is not a valid answer. Sample answer: because 1.43 is less than 2 and 0.71 is less than 1, so the answer cannot be more than 3.

16. Draw the decimal points on each number on the left side of the equation so that the difference is correct as shown.

$$21.4 - 2.3 = 19.1$$

17. A strip of cloth is 1.14 meters in length. A tailor can create 7 equal sized strips from this piece or 2 equal sized strips from this piece. Use the number lines below to model each option.

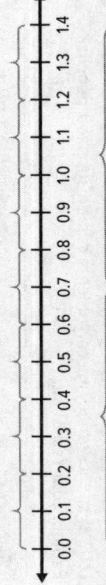

0.0 0.1 0.2 0.3 0.4 0.5 0.6 0.7 0.8 0.9 1.0 1.1 1.2 1.3 1.4

0.0 0.1 0.2 0.3 0.4 0.5 0.6 0.7 0.8 0.9 1.0 1.1 1.2 1.3 1.4

18. Aaron's work is shown for a recent operations on fractions test.

$$4 \div \frac{1}{4} = \frac{4 \div 1}{4} = \frac{4}{4} = 1$$

Part A: What mistake did Aaron make while dividing?

He divided the four by the 1 and then the 4, which is the same as dividing by 4, not $\frac{1}{4}$.

Part B: What is the correct quotient?

16

19. Explain how you model 0.7 × 0.8 using a decimal grid. How would this vary from 1.7 × 0.8?

Sample answer: For 0.7 × 0.8, I would use a 10 by 10 grid and shade 7 rows by 8 columns, which equals 0.56; For 1.7 × 0.8, I would show 10 rows by 8 columns and another 7 rows by 8 columns.

20. Which of the following uses the proper order of operations? Select all that apply.

- ▨ $3 \times (54 + 7) = 183$
- ☐ $12 - 5 + 2 = 5$
- ▨ $3 \times [5 - (6 \div 2)] = 6$
- ☐ $3^2 + 7 \times 4 = 37$
- ☐ $24 - (7 + 3 \times 2) = 40$

21. John is four more than 3 times his daughter's age.

Part A: If John's daughter is 10 years old, write an expression for John's age. Then calculate his age.

$3 \times 10 + 4 = 34$

Part B: If John's daughter is 9, how would John's age compare to your answer in *Part A*? Explain.

John would be three years younger, or 31 since the number multiplying the 3 would reduce by 1.

22. Draw lines between equivalent fractions.

$1\frac{3}{5}$ $\frac{13}{5}$

$2\frac{3}{5}$ $\frac{8}{5}$

$\frac{8}{32}$ $\frac{1}{4}$

$\frac{6}{32}$ $\frac{3}{16}$

23. A fire fighting robot needs to be programmed to go where the fire is. (0, 0) is considered the entrance to the room, and the fires are at points 2 right, 8 up; 2 up, 2 right; and 6 right, 2 up. Use the grid to draw and label the points where the fires are located.

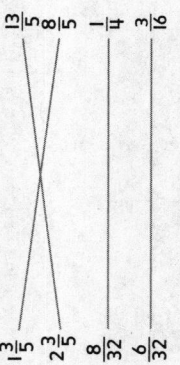

24. A tailor has a piece of cloth that is 1 yard wide by 1 yard long. From this he will cut a piece of cloth that is $\frac{5}{8}$ yard wide by $\frac{3}{4}$ yard long.

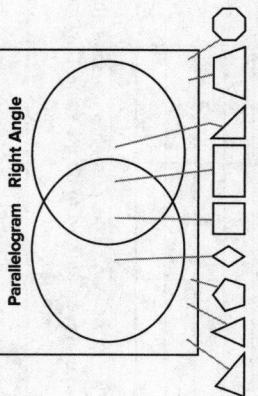

Part A: Model the desired cut on the square of fabric shown.

Part B: What is the area of the fabric? Write an equation.

$\frac{5}{8} \times \frac{3}{4} = \frac{15}{32}$ square yards

25. Use the Venn diagram to sort the shapes. Draw the line from the shape to the correct area of the diagram that describes it.

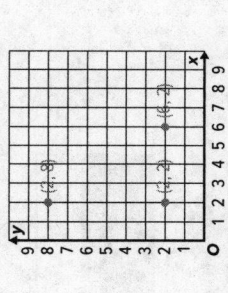

Parallelogram   Right Angle

26. Marcus is adding $\frac{2}{3}$ cup and $1\frac{1}{3}$ cups of cereal together in a bowl. In another bowl, he adds $\frac{5}{8}$ cup and $1\frac{1}{4}$ cups of dried fruit.

Part A: Find the sum of the contents of each bowl.

$\frac{2}{3} + 1\frac{1}{3} = 2$ cups; $\frac{5}{8} + 1\frac{1}{4} = \frac{5}{8} + 1\frac{2}{8} = 1\frac{7}{8}$ cups

Part B: Describe how the adding processes differed between finding the two sums.

The first set are like fractions, so I just needed to add the numerators. The second set I needed to find equivalent fractions before adding.

Part C: What is the total of contents in both bowls together?

$2 + 1\frac{7}{8} = 3\frac{7}{8}$ cups

27. The order for the Grade 5 class photo is shown. What is the total square footage of photo paper needed to print all the photos? Write an equation to show your work.

Grade 5 Class Photos

Quantity	Dimensions (in.)	Total Square Feet
7	$7\frac{3}{4} \times 5$	$38\frac{3}{4} \times 7 = 271\frac{1}{4}$
3	$3\frac{3}{8} \times 4$	$13\frac{1}{2} \times 3 = 40\frac{1}{2}$
8	$8 \times 10\frac{1}{2}$	$84 \times 8 = 672$

Total = $983\frac{3}{4}$ square feet

28. While measuring their bean plants after two weeks, the class recorded the following measurements in inches: $5\frac{1}{2}$, 6, $5\frac{1}{4}$, $6\frac{1}{4}$, $5\frac{1}{2}$, $6\frac{1}{4}$, 6, $5\frac{3}{4}$, $5\frac{3}{4}$

Part A: Use the line plot to record the measurements.

Heights of bean plants (inches)

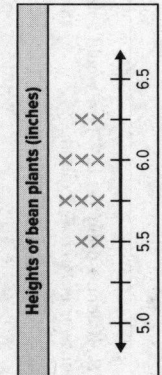

5.0   5.5   6.0   6.5

Part B: What is the total height of all the bean plants together?

$58\frac{3}{4}$ or 58.75 in.

29. You have a recipe for pancakes, but you want to increase the recipe so that it feeds $2\frac{1}{2}$ times the original number of people. How much of each ingredient do you need now?

Original Recipe	$2\frac{1}{2}$ Times Recipe
1 cup flour	$2\frac{1}{2}$ cups flour
$\frac{2}{3}$ cup sugar	$1\frac{2}{3}$ cups sugar
$\frac{3}{4}$ tablespoons baking powder	$1\frac{7}{8}$ tablespoons baking soda
$\frac{1}{2}$ teaspoon salt	$1\frac{1}{4}$ teaspoons salt

30. A cat jumped 3 times. Starting at the porch, he jumped 3 feet. Then he jumped 5 inches less than the first jump. Finally, he jumped $1\frac{1}{2}$ feet further than his second jump. What were the lengths of his jumps? How far from the porch is the cat now?

> Jump 1: 3 ft or 36 in.; Jump 2: 2 ft 7 in. or 31 in.: 36 in. − 5 in. = 31 in. or 2 ft 7 in.; Jump 3: 4 ft 1 in. or 49 in.: 2 ft 7 in. + 1 ft 6 in = 3 ft 13 in or 4 ft 1 in or 49 in.; 3 ft + 2 ft 7 in. + 4 ft 1 in. = 9 ft 8 in. or 116 in.

31. Kellen has 20 unit blocks. He needs to create rectangular prisms.

Part A: Use the table shown to create as many rectangular prisms as you can.

Length	Width	Height
1	1	20
1	2	10
1	4	5
2	2	5

Part B: As long as the prism has the same three numbers for the sides, it is considered to be the same. How many unique prisms can you create?

4

Page 187 • Building a Sandbox

Task Scenario Students will create a line plot and create a prism with fractional sides to model building a sandbox.		
Depth of Knowledge		DOK2, DOK3

Part	Maximum Points	Scoring Rubric
A	2	Full Credit: **Lengths of Boards (feet)** No credit will be given for an incorrect answer.
B	3	Full Credit: 42 ft **57,024 cubic inches** Partial Credit (1 point) will be given for the correct board length **OR** the correct construction of the prism **OR** the correct volume. No credit will be given for an incorrect answer.
C	2	Full Credit: Sample answer: The new volume will be half of the volume in **Part B**. Since the old volume is $6\ell h$ and the new volume is $3\ell h$, the new volume will be half the size of the old volume. Partial Credit (1 point) will be given for correctly stating the relationship between the volumes without a correct explanation. No credit will be given for an incorrect answer.
TOTAL	7	